Building Construction and Planning Handbook

Building Construction and Planning Handbook

Edited by **Seth Royal**

CLANRYE
INTERNATIONAL

New Jersey

Published by Clanrye International,
55 Van Reypen Street,
Jersey City, NJ 07306, USA
www.clanryeinternational.com

Building Construction and Planning Handbook
Edited by Seth Royal

International Standard Book Number: 978-1-63240-090-1 (Hardback)

Printed in the United States of America.

Contents

Preface

The creation of any new product requires planning. Building construction, as the name suggests, essentially encompasses all the stages and aspects which are pre-requisites for constructing a new building. This includes preparing plans, designing, forming and financing construction of buildings, till the stage when the structure is ready for occupancy.

Such projects usually begin with a conceptual estimate performed by a building estimator. This is followed by highly complex financial plans and a complex net of contracts and other legal obligations, which must be carefully considered. The planning and designing process begins with a series of drawings and specifications. Numerous experts from various fields such as quantity estimators, surveyors, civil engineers, archaeological consultants and fire protection engineers are involved in finally determining the overall layout of the building.

Adhering to the legal framework governing the property is another crucial aspect of planning. This includes following governmental regulations related to property laws, adherence to the local building code and other obligations that are created in the process of construction. Once the construction is complete, a final inspection is made after which an occupancy permit is issued.

Clearly, building construction and planning is a complex procedure which involves comprehensive knowledge. This book is an attempt to provide an insight to its readers on the various nuances in this discipline. I thank all the contributors who've shared their studies with us for this book. I would also like to thank my family for their constant support and faith in me.

Editor

Comparison of School Building Construction Costs Estimation Methods Using Regression Analysis, Neural Network, and Support Vector Machine

Gwang-Hee Kim[1], Jae-Min Shin[2], Sangyong Kim[3], Yoonseok Shin[1*]

[1]Department of Plant & Architectural Engineering, Kyonggi University, Suwon-Si, Korea; [2]Department of Architectural Engineering, Graduate School of Kyonggi University, Suwon-Si, Korea; [3]School of Construction Management and Engineering, University of Reading, Reading, UK.

ABSTRACT

Accurate cost estimation at the early stage of a construction project is key factor in a project's success. But it is difficult to quickly and accurately estimate construction costs at the planning stage, when drawings, documentation and the like are still incomplete. As such, various techniques have been applied to accurately estimate construction costs at an early stage, when project information is limited. While the various techniques have their pros and cons, there has been little effort made to determine the best technique in terms of cost estimating performance. The objective of this research is to compare the accuracy of three estimating techniques (regression analysis (RA), neural network (NN), and support vector machine techniques (SVM)) by performing estimations of construction costs. By comparing the accuracy of these techniques using historical cost data, it was found that NN model showed more accurate estimation results than the RA and SVM models. Consequently, it is determined that NN model is most suitable for estimating the cost of school building projects.

Keywords: Estimating Construction Costs; Regression Analysis; Neural Network; Support Vector Machine

1. Introduction

In school building construction projects, budgeting, planning, and monitoring for compliance with the client's available budget, time, and work outstanding are also important [1]. And the accuracy of construction costs estimation is a key factor in the success of a construction project, and also affects the decision-making by the owners [2-4]. But it is difficult to quickly and accurately estimate the construction costs at the planning stage, because the drawings and documentation are generally incomplete [5]. For this reason, various techniques have been developed to accurately estimate construction costs with the limited project information available in the early stage.

Typical cost estimating techniques are neural networks (NN), support vector machine (SVM), case-based reasoning (CBR), and regression analysis (RA), etc. [6]. For example, the RA model [7-9], NN model [10-13], SVM Model [6,14-16], and CBR model [1,17,18] have been developed for predicting or estimating construction costs. Approaches to cost estimation based on statistics and linear regression analysis have been developed since the 1970s [2]. Since the late 1980s, artificial intelligence

approaches such as expert system, NN, and CBR have been applied [19]. In addition, the cost predicting model has been studied since the 2000s.

Previous studies [2,12,20-22] revealed that an NN model for cost estimating is superior to the RA model. Also, the accuracy of cost estimation based on the SVM technique is similar to that of cost estimation based on RA [23]. Consequently, it is necessary to compare RA, NN, and SVM to determine the optimum approach to estimating construction costs.

Therefore, in this research, the accuracy of three estimating techniques (*i.e.* regression analysis, neural network, and support vector machine techniques) is compared by performing an estimation of construction costs using historical cost data, so that a cost estimation model adapting two techniques (*i.e.* neural network and support vector machine) could be examined through regression analysis.

2. Three Costs Estimating Techniques

2.1. Regression Analysis

Some studies have mentioned that cost estimating models

using regression analysis have several disadvantages: 1) they have no specific, or clearly defined, approach that will help estimators choose the cost model that best fits the historical data to a given cost estimating application [12,20,24,25]; 2) a certain type of multiple equation and its data are assumed to be similar to be suitable for the regression equation [12,24,25]; 3) the variable influencing the estimation must be reviewed in advance, and it is also difficult to use a large number of input variables [24-26]. However, regression analysis, as it is usually called, is a very powerful statistical tool that can be used as both an analytical and predictive technique in examining the contribution of potential new items to the overall estimate reliability [27]. Regression analysis (RA) can be generally represented the form of Equation (1).

$$Y = C + A_1X_1 + A_2X_2 + \cdots\cdots + A_nX_n \qquad (1)$$

where Y is the total estimated costs, and X_1, X_2, $\cdots X_n$ are measures of distinguishable variables that may help in estimating Y, C is the estimated constant, and A_1, A_2, $\cdots$, A_n are coefficients estimated by regression analysis, given the availability of some relevant data. The Statistical Package for Social Science (SPSS) stepwise techniques were used to develop the regression model.

2.2. Neural Network

A neural network (NN) is a computer system that simulates the learning process of the human brain [2] based on a simplified model of the biological neurons in the human brain and the relations between them. A neural network is modeled in a mathematical manner to implement an intelligent form as shown in the human brain, for utilization in engineering or in other fields [3]. The structure of an NN is as shown in **Figure 1**. Basically, the network consists of several layers, including an input layer, a hidden layer, and an output layer, and each layer contains neurons. Neurons determine the optimum value through a summation and transfer function. The set of inputs, which is the outputs from another neuron in input layers, are delivered by neurons. Each input data is multiplied by the connection weight, and then the weighted inputs provide output value, which is modified by the transfer function.

Some researchers have explored the application of NN to improve the accuracy of cost estimation beyond that of the regression model [10-12,20,24,25,28,29]. Although many previous studies have proved that the neural network cost estimating model is superior to the regression analysis estimation model, many have also demonstrated not only the superiority of NN but the problems associated with using them for cost estimation [4]. However, the main advantages of an NN are as follows: 1) they can be used to construct high-level nonlinear function esti-

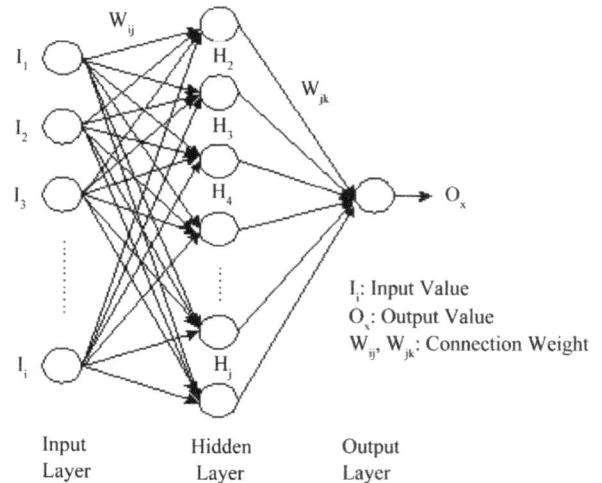

Figure 1. Neural network structure.

mation models; and 2) their use does not impose any limit on the number of features [30]. The main disadvantage of the NN mentioned in the previous research is that the black box techniques and knowledge acquisition process are very time-consuming [11,28,29,31].

2.3. Support Vector Machine

Support vector machine (SVM) is a learning theory developed by Vapnik [32] that has two main categories, support vector classification (SVC) and support vector regression (SVR). In particular, in the model constructed using SVR, the goal is to find a function $f(x)$ that has at most ε deviation from the actually obtained target value (y_i) for all the training data, and is simultaneously as flat as possible [33]. The structure of SVR is as shown in **Figure 2**. The input pattern (support vectors) is mapped into feature space by a map Φ. Then, dot products are computed with the images of the training patterns under the map Φ. This matches up to the evaluating kernel function $k(x_i, x)$. The dot products are aggregated using the weights $v_i = \alpha_i - \alpha_i^*$. Last, the final prediction output is calculated by adding the constant value (b).

In most cases, the performance of SVM generation either matches or is significantly better than that of competing methods such as NN and fuzzy system [34]. However, by comparison with NN, research to apply SVM to cost estimation has not yet been active. Therefore, with only a few studies [6,16,23,35], it is too early to conclude that SVM has superiority in cost estimation. The main advantage of SVM is the capacity for self-learning and high performance in generalization [36]. The main disadvantages of SVM are as follows: 1) it requires a trial and error period to determine both a suitable kernel function and the parameters of the kernel function [16]; 2) SVM models have a high level of algorithmic

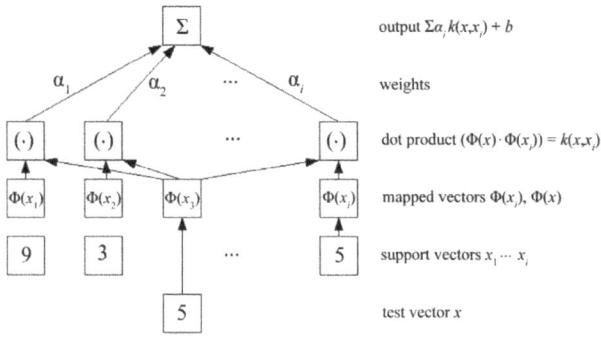

Figure 2. Support vector regression structure.

complexity and require extensive memory [37].

3. Application

3.1. Data for Estimating Construction Costs

The collected data used in this application were the actual construction costs of 217 school building projects executed by general contractors from 2004 to 2007 in Kyeonggi Province, Korea. These cost data were the direct costs of school buildings, such as elementary, middle, and high schools, without mark-up. As shown in **Table 1**, 10 input and 1 output variables were extracted from the collected data. Notably, the construction year was not used as an input variable because the extracted variables from cost data were converted using the Korean building cost index (BCI), *i.e.* the collected cost data were multiplied by the BCI of the base year 2005 (BCI = 1.00). The collected cost data of 217 school buildings were divided randomly into 20 test data, 67 cross-validation data, and 130 training data.

3.2. Accuracy Evaluation

Generally, the performance of a cost estimating model is determined by measuring its bias, consistency, and accuracy. Measures of bias, consistency, and accuracy are concerned with the difference in the average between the actual costs and the estimated costs, considering both the degree of variation around the average, and the combination with bias and consistency [2]. By far, the most popular evaluation criteria used involves statistics such as mean, standard deviation, and coefficient of variation [38]. In this research, each model's performance was measured by the Mean Absolute Error Rates (MAERs), which was calculated by Equation (2).

$$\text{MAERs} = \frac{\left(\sum \left| \frac{C_e - C_a}{C_a} \times 100 \right| \right)}{n} \quad (2)$$

where C_e is the estimated construction costs by model application, C_a is the collected actual construction costs,

Table 1. Input and output variables.

	Description	Min	Max	Average	Remark
	Year	From 2004 to 2007			None
	Budget	1. BTL 2. National Finance			Nom.
	School Levels	1. Elementary 2. Middle 3. High			Nom.
Input	Land Acquisition	1. Existing 2. Building lots 3. Green Belts			Nom.
	Class Number	12	48	31	Num.
	Building Area	1204	3863	2694	Num.
	Gross Floor Area	4925	12,710	9656	Num.
	Storey	3	7	4.7	Num.
	Basement Floor	0	2	0.5	Num.
	Floor Height	3.3	3.6	3.5	Num.
Output	Total Construction Cost				

and n is the number of test data.

3.3. Results of Evaluation

The results from the 20 test data using RA, NN and SVM are summarized in **Tables 2** and **3**. The results from the RA model had MAERs of 5.68 with 20% of the estimates within 2.5% of the actual error rate, while 80% were within 10%. The NN model had MAERs of 5.27 with 35% of the estimates within 2.5% of the actual error rate, while 85% were within 10%. Last, SVM model had MAERs of 7.48 with 10% of the estimates within 2.5% of the actual error rate, while 75% were within 10%. Also, the standard deviation of the RA, NN, and SVM model are 3.56, 4.13, and 4.66 respectively, as shown in **Table 4** and **Figure 3**.

4. Discussion of Results

This study was conducted by using 217 cases of school building construction projects. Of the cases, 20 cases were used for the testing. The regression model, neural networks model, and support vector machine model with 20 test data gave MAERs of 5.68, 5.27 and 7.48, respectively. Also, the NN model and the RA model had smaller error rates and deviation than that of SVM model. Through the performance, the NN model was the most accurate and reliable of the three models.

The MAERs of three results were then compared using analysis of variance (ANOVA). The MAERs of three results would be statistically similar, even if there were differences between them. The null hypothesis is that MAERs of the three results are all equal ($H_0 : u_D = 0$). The F-statistic is the ratio of the mean squares between the variance of three results. If the F ratio is close to "1", the null hypothesis is rejected. This analysis showed that

Table 2. Summarized results by estimating model.

Error rate (%)	RA		NN		SVM	
	Fre. (%)	Cum. (%)	Fre. (%)	Cum. (%)	Fre. (%)	Cum. (%)
0.0 - 2.5	4 (20)	4 (20)	7 (35)	7 (35)	2 (10)	2 (10)
2.5 - 5.0	6 (30)	10 (50)	3 (15)	10 (50)	6 (30)	8 (40)
5.0 - 7.5	3 (15)	13 (65)	4 (20)	14 (70)	1 (5)	9 (45)
7.5 - 10.0	3 (15)	16 (80)	3 (15)	17 (85)	6 (30)	15 (75)
10.0 - 12.5	4 (20)	20 (100)	2 (10)	19 (95)	3 (15)	18 (90)
12.5 - 15.0	0 (0)	20 (100)	1 (5)	20 (100)	0 (0)	18 (90)
15.0 - 17.5	0 (0)	20 (100)	0 (0)	20 (100)	2 (10)	20 (100)
MAERs	5.68	-	5.27	-	7.48	-

Table 3. Results of estimating costs of each test set.

No,	Historical Cost (1000 KRW)	Regression Analysis		Neural Networks		Support Vector Machine	
		Predicted Cost	Error Rate (%)	Predicted Cost	Error Rate (%)	Predicted Cost	Error Rate (%)
1	6,646,893	5,941,793	10.61	6,289,870	5.37	6,088,979	8.39
2	7,156,246	7,420,364	3.69	7,584,598	5.99	7,417,214	3.65
3	8,148,347	8,881,124	8.99	9,088,620	11.54	8,109,849	0.47
4	9,084,489	9,339,891	2.81	9,649,732	6.22	8,832,427	2.77
5	8,806,657	8,529,327	3.15	8,710,677	1.09	8,387,286	4.76
6	7,878,570	7,883,956	0.07	7,701,203	2.25	7,764,850	1.44
7	6,770,824	7,256,146	7.17	7,448,678	10.01	7,553,526	11.56
8	9,208,789	8,809,574	4.34	9,167,173	0.45	8,457,476	8.16
9	10,690,800	9,803,464	8.30	10,318,258	3.48	9,362,775	12.42
10	7,605,493	8,481,519	11.52	8,609,664	13.20	7,907,033	3.96
11	7,764,941	7,490,563	3.53	7,682,920	1.06	8,385,352	7.99
12	6,860,022	7,092,590	3.39	6,285,265	8.38	6,223,925	9.27
13	7,571,841	7,711,195	1.84	7,531,603	0.53	8,125,465	7.31
14	10,801,826	9,575,984	11.35	9,749,778	9.74	9,022,620	16.47
15	9,259,063	9,052,071	2.24	9,269,961	0.12	8,485,674	8.35
16	7,966,421	8,406,525	5.52	8,286,950	4.02	7,300,706	8.36
17	6,777,067	6,260,888	7.62	6,764,268	0.19	7,013,327	3.49
18	6,895,476	6,996,417	1.46	7,226,609	4.80	8,088,323	17.30
19	7,478,649	7,905,854	5.71	8,037,224	7.47	8,269,068	10.57
20	6,779,866	7,475,533	10.26	7,422,840	9.48	6,971,690	2.83
MAERs			5.68		5.27		7.48

Table 4. Descriptive analysis of estimating error rate.

	Mean	Std, deviation	Std, error	95% confidence interval of the mean	
				Lower	Upper
RA	5.6785	3.56508	.79718	4.0100	7.3470
NN	5.2695	4.13996	.92572	3.3319	7.2071
SVM	7.4760	4.66776	1.04374	5.2914	9.6606

Comparison of School Building Construction Costs Estimation Methods Using Regression Analysis, Neural
Network, and Support Vector Machine

5

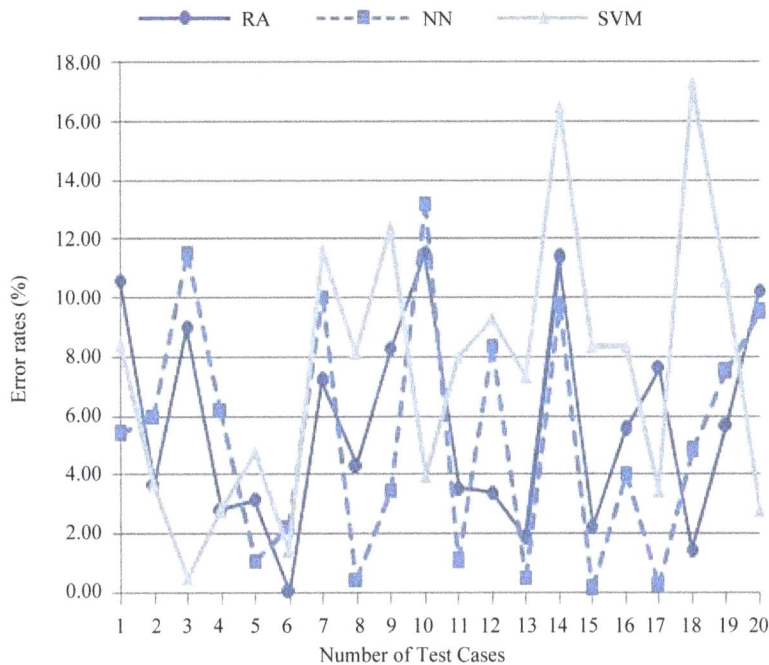

Figure 3. Comparison of the results of each model.

the MAERs of the three results were statistically different. Therefore, the NN model performed more effectively than the other two results in estimating construction costs.

5. Conclusions

This study applied the three techniques of RA, NN, and SVM to estimate the construction cost of school building projects. 197 cases were used for model development and validation, while the remaining 20 cases were used for testing the model. All three models produced a high correlation between the estimating costs and the actual costs.

Although RA, NN, and SVM worked well for the application, NN model gave more accurate estimation results than the RA and SVM models. As mentioned in the previous research, NN has proven to be useful and suitable for dealing with complex problems and developing user-friendly predictive models. They are able to detect any patterns found in the data and provide a greater opportunity to investigate different options and project control techniques. Also, in this study, the NN estimating model is more suitable for estimating school building projects than the SVM estimating model.

REFERENCES

[1] G.-H. Kim, J.-E. Yoon, S.-H. An, H-H. Cho and K.-I. Kang, "Neural Network Model Incorporating a Genetic Algorithm in Estimating Construction Costs," *Building and Environment*, Vol. 39, No. 11, 2004, pp. 1333-1340.

[2] G.-H. Kim and S.-H. An, "A Study on the Correlation between Selection Methods of Input Variables and Number of Data in Estimating Accuracy; Cost Estimating Using Neural Networks in Apartment Housing Projects," *Journal of the Architectural Institute of Korea*, Vol. 23, No. 4, 2007, pp. 129-137.

[3] G.-H. Kim, S.-H. An and K.-I. Kang, "Comparison of Construction Cost Estimating Models Based on Regression Analysis, Neural Networks, and Case-Based Reasoning," *Building and Environment*, Vol. 39, No. 10, 2004, pp. 1235-1242.

[4] H.-G. Cho, K.-G. Kim, J.-Y. Kim and G.-H. Kim, "A Comparison of Construction Cost Estimation Using Multiple Regression Analysis and Neural Network in Elementary School Project," *Journal of the Korea Institute of Building Construction*, Vol. 13, No. 1, 2013, pp. 66-74.

[5] S.-H. An and K.-I. Kang, "A Study on Predicting Construction Cost of Apartment Housing Using Experts' Knowledge at the Early Stage of Projects," *Journal of the Architectural Institute of Korea*, Vol. 21, No. 6, 2005, pp. 81-88.

[6] U.-Y. Park and G.-H. Kim, "A Study on Predicting Construction Cost of Apartment Housing Projects Based on Support Vector Regression at the Early Project Stage," *Journal of the Architectural Institute of Korea*, Vol. 23, No. 4, 2007, pp. 165-172.

[7] S. Singh, "Cost Model for Reinforced Concrete Beam and Slab Structures in Building," *Journal of Construction Engineering and Management*, Vol. 116, No. 1, 1990, pp. 54-67.

[8] K.-D. Kim, "A Study on the Development of the Cost

Model for the Domestic Apartment House," Ph.D. Thesis, Seoul National University, Seoul, 1991.

[9]   I.-S. Choi, S.-H. Hong, C.-B. Son and S.-C. Ko, "A Study on the Prediction Model of Construction Cost in High-Rise Office Building of SRC Type," *Journal of the Architectural Institute of Korea*, Vol. 15, No. 7, 1999, pp. 143-151.

[10]   R. Mckim, "Neural Network Application to Cost Engineering," *Cost Engineering*, Vol. 35, No. 7, 1993, pp. 31-35.

[11]   I.-C. Yeh, "Quantity Estimating of Building with Logarithm-Neuron Networks," *Journal of Construction Engineering and Management*, Vol. 124, No. 5, 1998, pp. 374-380.

[12]   J. Bode, "Neural Networks for Cost Estimating: Simulation and Pilot Application," *International Journal of Production Research*, Vol. 38, No. 6, 2000, pp. 123-154.

[13]   S.-K. Kim and I.-W. Koo, "A Neural Network Cost Model for Office Buildings," *Journal of the Architectural Institute of Korea*, Vol. 16, No. 9, 2000, pp. 59-67.

[14]   X. Wu and L. Cai, "Application of RS-SVM in Construction Project Cost Forecasting," *Proceedings of the 4th International Conference on Wireless Communication, Networking and Mobile Computing*, Dalian, 12-14 October 2008, pp. 1-4.

[15]   M.-Y. Cheng and Y.-W. Wu, "Construction Conceptual Cost Estimates Using Support Vector Machine," *Proceedings of the 22nd International Symposium on Automation and Robotics in Construction ISARC 2005*, Ferrara, 11-14 September 2005, pp. 1-5.

[16]   S.-H. An, K.-I. Kang, M.-Y. Cho and H.-H. Cho, "Application of Support Vector Machines in Assessing Conceptual Cost Estimates," *Journal of Computing in Civil Engineering*, Vol. 21, No. 4, 2007, pp. 259-264.

[17]   W. Yunna, "Application of a Case-Based Reasoning Method in Estimating the Power Grid Project Cost," *Proceedings of the 4th International Conference on Wireless Communication*, Networking and Mobile Computing, Dalian, 12-14 October 2008, pp. 1-5.

[18]   S.-H. Ji, M. Park and H.-S. Lee, "Case Adaptation Method of Case-Based Reasoning for Construction Cost Estimation in Korea," *Journal of Construction Engineering and Management*, Vol. 138, No. 1, 2007, pp. 43-52.

[19]   S.-H. An, G.-H. Kim and K.-I. Kang, "A Case-Based Reasoning Cost Estimating Model Using Experience by Analytic Hierarchy Process," *Building and Environment*, Vol. 42, No. 7, 2007, pp. 2573-2579.

[20]   J. Garza and K. Rouhana, "Neural Network versus Parameter-Based Application," *Cost Engineering*, Vol. 37, No. 2, 1995, pp. 14-18.

[21]   W.-Y. Park, J.-H. Cha and K.-I. Kang, "A Neural Network Cost Model for Apartment Housing Projects in the Initial Stage," *Journal of the Architectural Institute of Korea*, Vol. 18, No. 7, 2002, pp. 155-162.

[22]   G.-H. Kim, S.-H. An and H.-K. Cho, "Comparison of the Accuracy between Cost Prediction Models Based on Neural Network and Genetic Algorithm: Focused on Apartment Housing Project Cost," *Journal of the Architectural Institute of Korea*, Vol. 23, No. 3, 2006, pp. 111-118.

[23]   J.-M. Shin and G.-H. Kim, "A Study on Predicting Construction Cost of Educational Building Project at Early Stage Using Support Vector Machine Technique," *Journal of Korean Institute of Educational Environment*, Vol. 11, No. 3, 2012, pp. 46-54.

[24]   H. Adeli and M. Wu, "Regularization Neural Network for Construction Cost Estimation," *Journal of Construction Engineering and Management*, Vol. 124, No. 1, 1998, pp. 18-24.

[25]   J. Bode, "Neural Networks for Cost Estimation," *Cost Engineering*, Vol. 40, No. 1, 1998, pp. 25-30.

[26]   A. E. Smith and A. K. Mason, "Cost Estimating Predictive Modeling: Regression versus Neural Network," *Engineering Economist*, Vol. 42, No. 2, 1997, pp. 137-161.

[27]   R. M. Skitmore and B. R. T. Patchell, "Development in Contract Price Forecasting and Bidding Techniques," In: M. Skitmore and V. Marston, Eds., *Cost Modelling*, E& FN Spon, London, 1990, pp. 53-84.

[28]   R. Creese and L. Li, "Cost Estimation of Timber Bridge Using Neural Networks," *Cost Engineering*, Vol. 37, No. 5, 1995, pp. 17-22.

[29]   H. Li, "Neural Networks for Construction Cost Estimation," *Building Research and Information*, Vol. 23, No. 5, 1995, pp. 279-284.

[30]   S. Deng and T.-H. Yeh, "Applying Least Squares Support Vector Machines to the Airframe Wing-Box Structural Design Cost Estimation," *Expert Systems with Applications*, Vol. 37, No. 12, 2010, pp. 8417-8423.

[31]   T. Hegazy, P. Fazio and O. Moselhi, "Developing Practical Neural Network Application Using Back-Propagation," *Computer-Aided Civil and Infrastructure Engineering*, Vol. 9, No. 2, 1994, pp. 145-159.

[32]   V. N. Vapnik, "The Nature of Statistical Learning Theory," Springer, London, 1999.

[33]   A. J. Smola and B. Schölkopf, "A Tutorial on Support Vector Regression," *Statistics and Computing*, Vol. 14, No. 3, 2004, pp. 199-222.

[34]   C. J. C. Burges, "A Tutorial on Support Vector Machines for Pattern Recognition," *Data Mining and Knowledge Discovery*, Vol. 2, No. 2, 1998, pp. 121-167.

[35]   M.-Y. Cheng, H.-S. Peng, Y.-W. Wu and T.-L. Chen, "Estimate at Completion for Construction Projects Using Evolutionary Support Vector Machine Inference Model," *Automation in Construction*, Vol. 19, No. 5, 2010, pp.

Comparison of School Building Construction Costs Estimation Methods Using Regression Analysis, Neural Network, and Support Vector Machine

7

619-629.

[36] Y. Shin, D.-W. Kim, J.-Y. Kim, K.-I. Kang, M.-Y. Cho and H.-H. Cho, "Application of Adaboost to the Retaining Wall Method Selection in Construction," *Journal of Computing in Civil Engineering*, Vol. 23, No. 3, 2009, pp. 188-192.

[37] P. R. Kumar and V. Ravi, "Bankruptcy Prediction in Banks and Firms via Statistical and Intelligent Techniques—A Review," *European Journal of Operational Research*, Vol. 180, No. 1, 2007, pp. 1-28.

[38] M. Skitmore, "Early Stage Construction Price Forecasting: A Review of Performance," Occasional Paper, Royal Institute of Chartered Surveyors, London, 1991.

2

Tests on Alkali-Activated Slag Foamed Concrete with Various Water-Binder Ratios and Substitution Levels of Fly Ash

Keun-Hyeok Yang[1*], Kyung-Ho Lee[2]

[1]Department of Plant & Architectural Engineering, Kyonggi University, Suwon, Korea; [2]Department of Architectural Engineering, Kyonggi University Graduate School, Seoul, Korea.

ABSTRACT

To provide basic data for the reasonable mixing design of the alkali-activated (AA) foamed concrete as a thermal insulation material for a floor heating system, 9 concrete mixes with a targeted dry density less than 400 kg/m^3 were tested. Ground granulated blast-furnace slag (GGBS) as a source material was activated by the following two types of alkali activators: 10% $Ca(OH)_2$ and 4% $Mg(NO_3)_2$, and 2.5% $Ca(OH)_2$ and 6.5% Na_2SiO_3. The main test parameters were water-to-binder (W/B) ratio and the substitution level (R_{FA}) of fly ash (FA) for GGBS. Test results revealed that the dry density of AA GGBS foamed concrete was independent of the W/B ratio an R_{FA}, whereas the compressive strength increased with the decrease in W/B ratio and with the increase in R_{FA} up to 15%, beyond which it decreased. With the increase in the W/B ratio, the amount of macro capillaries and artificial air pores increased, which resulted in the decrease of compressive strength. The magnitude of the environmental loads of the AA GGBS foamed concrete is independent of the W/B ratio and R_{FA}. The largest reduction percentage was found in the photochemical oxidation potential, being more than 99%. The reduction percentage was 87% - 93% for the global warming potential, 81% - 84% for abiotic depletion, 79% - 84% for acidification potential, 77% - 85% for eutrophication potential, and 73% - 83% for human toxicity potential. Ultimately, this study proved that the developed AA GGBS foamed concrete has a considerable promise as a sustainable construction material for nonstructural element.

Keywords: Alkali-Activated Foamed Concrete; Granulated Ground Blast-Furnace Slag; Fly Ash; Water-to-Binder Ratio; Environmental Load

1. Introduction

Most of residential buildings and houses in Korea adopt a floor heating system. The floor radiation heating system is commonly estimated to be lower as much as 50% in energy consumption compared with convection heating system [1]. Furthermore, the floor heating system can be converted to a floor cooling system by using a heat pump system. As a result, several countries have been recently interested in the floor heating system to overcome the limitation of convection heating system and enhance indoor environmental quality and comfort including energy conservation. In floor heating system, foamed concrete is commonly constructed between reinforced concrete slab and finishing mortar covering heating pipes to minimize a heat loss through concrete slab and to maintain the layout of heating pipe during construction of finishing mortar, as shown in **Figure 1**. Hence, foamed concrete for floor heating system fundamentally requires prefera-

bly lower density for lighter self-weight and lower thermal conductivity, while it also needs a minimum compressive strength at an early age to fix heating pipes and prevent its bearing failure during construction of heating pipe and finishing mortar.

In recent, various efforts have been attempted to reduce the use of ordinary Portland cement (OPC) for concrete production, because the OPC is generally estimated to be predominantly responsible for the environmental loads in the concrete industry [2,3]. One of the active alternatives for the OPC, an alkali-activated (AA) binder has begun to attract a great concern since the late 1980's, though further investigation and experimental verification on various essential performances including mechanical properties, inelastic behavior and durability need for the AA binder to practically apply to structural members. However, several reviews [4-8] reveal that the AA binder can draw a good strength gain property and beneficial environmental impact with low CO_2 emission

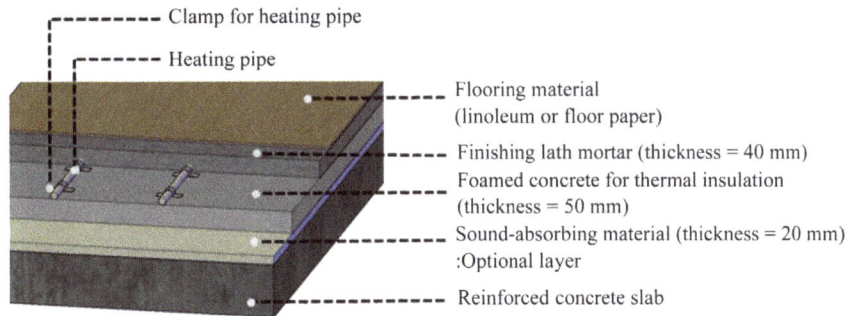

Figure 1. Typical section details of floor heating system.

and recycling of by-product materials such as ground granulated blast-furnace slag (GGBS) and fly ash (FA). Hence, AA binder would be considerable promise to produce a sustainable non-structural concrete element. Esmaily and Nuranian [9] showed that GGBS activated by sodium silicate solution can produce non-autoclaved high strength cellular concrete when the ratio of sodium silicate solution to aluminum powder for gas production is close to 1.0. Yang *et al.* [10] tested to develop AA GGBS foamed concrete as thermal insulation material for floor heating system and concluded that the unit binder content of approximately 400 kg/m^3 is required to achieve the minimum quality requirements specified in KS F 4039 [11] and ensuring the economic efficiency. On the other hand, the air-void system of foamed concrete, which significantly influences the strength and thermal conductivity of such concrete, depends on mixing proportions including water-to-binder ratio, foam volume and amount of filler as well as the performance of foam agent [12]. Furthermore, another current concern for the foamed concrete is sustainable quality. Therefore, various experiments and statistical data banks are required to establish a reliable mixing proportion for AA GGBS foamed concrete with a lower environmental impact.

The present experimental program was conducted as a follow-up to the previous investigation [10] to suggest a reasonable mixing proportion of AA GGBS foamed concrete for thermal insulation material in a floor heating system based on the various test data and understanding the behavior of such concrete. The main test parameters were water-to-binder (W/B) ratio and the substitution level (R_{FA}) of FA for GGBS. The quality and availability of the mixed foamed concrete were examined through comparisons with the minimum requirements (initial flow and defoamed depth for fresh concrete, and dry-density and compressive strength for hardened concrete) specified in KS F 4309 [11]. The effects of W/B ratio and RF on the characteristics of the air-void structure of the hardened concrete were ascertained by mercury intrusion porosimetry and light optical microscope. The environmental impacts of test mixes were also compared with

those calculated from a typical OPC foamed concrete mix.

2. Experimental Details

2.1. Specimens and Mixing Proportions

All specimens were classified into two groups according to the selected test parameters, as given in **Table 1**. The mixing details for specimens of Group I are as follows: water-to-binder (W/B) ratio as the main parameter varied from 40% to 50%; a combination of 10% Ca(OH)$_2$ and 4% Mg(NO$_3$)$_2$ was used for an activator; and the unit binder content was fixed at 375 kg/m^3. It is noted that the binder includes the source materials and activators. For specimens of Group II, FA was substituted for GGBS with the range of 0% - 20%, and then those source materials were activated by 2.5% Ca(OH)$_2$ and 6.5% Na$_2$SiO$_3$. The unit binder content and W/B ratio for mixes of Group II were fixed at 425 kg/m^3 and 37.5%, respectively. The foamed concrete for a floor heating system practically requires high workability with flow above 180 mm for self-compactability [1]. To achieve the high workability of all mixes, high early-strength agent containing water-reducing efficiency was added for mixes of Group I by 0.2% binder content, while 0.75% naphthalene-based high-range water-reducing agent was added for mixes of Group II. The foam volume required for a given unit binder content and W/B ratio was determined based on the procedure specified in ASTM C796-97 [13] together with the unit volume of concrete.

2.2. Materials

The chemical compositions of GGBS and FA obtained from x-ray fluorescence (XRF) analysis are given in **Table 2**. The GGBS was mainly composed of calcium, silicon, alumina and magnesium oxides. The FA used had a low calcium oxide (CaO) but was rich in both silicon and alumina as the silicon oxide (SiO$_2$)-to-aluminum oxide (Al$_2$O$_3$) ratio by mass is 1.91 which belongs to class F. The Blaine fineness and specific gravity were 4400 cm^2/g and 2.9, respectively, for GGBS, and 4200 cm^2/g

Table 1. Mixing proportions of the foamed concrete specimens and summary of test results.

Group	Specimens	Designed foam volume ratio (%)	W/B ratio (%)	R_{FA} (%)	Unit binder content (kg/m³)	GGBS	FA	Ca(OH)₂ (%)	Na₂SiO₃ (%)	Mg(NO₃)₂ (%)	Actual foam volume ratio (%)	Flow (mm)	Defoamed depth (mm)	Dry density (kg/m³)	7 days	28 days
I	I-40	71.24	40								72.5	≥250	0	375	2.15	3.03
	I-45	69.37	45								67.5	≥250	0	380	1.26	1.93
	I-47.5	68.43	47.5	0	375	86	-	10	-	4	67.1	≥250	0	391	1.06	1.63
	I-50	67.49	50								65.5	≥250	0	405	0.95	1.53
II	II-0	69.00		0		91	-				69.3	195	11	406	0.83	1.20
	II-5	68.76		5		86	5				68.1	200	12	409	0.97	1.68
	II-10	68.53	37.5	10	425	81	10	2.5	6.5	-	67.8	215	14	403	1.54	2.16
	II-15	68.30		15		76	15				67.5	215	16	404	1.39	1.67
	II-20	68.06		20		71	20				67.8	220	17	405	0.87	1.30

Note: In specimen notations, the first and second parts indicate the affiliated group and main parameter in each group, respectively. Hence, the second parts for Groups I and II refer to water-to-binder ratio and substitution level of FA for GGBS, respectively. For example, specimen I-40 indicates the foamed concrete mix with W/B ratio of 40% being affiliated to Group I, while specimen II-5 indicates the foamed concrete mix with R_{FA} of 5% being affiliated to Group II. *Concrete mixes of Group I contain high early-strength agent with water reducing efficiency by 0.2% binder content; **Concrete mixes of Group II contain 0.75% naphthalene-based high-range water-reducing admixture.

Table 2. Chemical composition of selected source materials (% by mass).

Materials	SiO₂	Al₂O₃	Fe₂O₃	CaO	MgO	K₂O	Na₂O	TiO₂	SO₃	LOI*
FA	57.70	28.60	5.08	4.70	0.67	0.57	0.37	1.53	0.68	0.1
GGBS	34.70	13.80	0.11	44.60	4.38	0.48	-	0.74	0.95	0.24

*Loss on ignition.

and 2.2, respectively, for FA. All dry powdered alkali activators were pre-blended with source materials in the dry form. The specific gravity and maximum particle sizes were 2.24 and 21.2 mm, respectively, for Ca(OH)₂, 2.2 and 1026.1 mm, respectively, for Na₂SiO₃, 1.56 and 600 mm, respectively, for Mg(NO₃)₂. The foaming agent used to produce pre-foamed foam was a vegetable-based soap-type resin that is generally applied for OPC foamed concrete.

2.3. Mixing and Testing

All concrete specimens were produced in accordance with the pre-foamed foam mixing procedure recommended in ASTM C796-97 [13]. To produce foam, the foaming agent was diluted with water in the ratio of 1:19 by volume and then aerated to a density of 40 kg/m³ using a foam generator. The pre-formed foam was added to cementitious slurry and then mixed in a 0.12-m³ capacity circulating mixer pan.

For the fresh concrete, the initial flow tested without raising and dropping the flow table and defoamed depth were recorded in accordance with KS F 4039 [11]. The actual foam volume in the fresh concrete was also measured using a mess cylinder and methyl alcohol, with reference to the method proposed by Lee *et al.* [14]. The

compressive strength of concrete was measured using 100 × 200 mm cylinder at ages of 7 and 28 days. The dry density of hardened concrete was measured in accordance with KS F 2459 [11]. The pore size distribution and air-void structure were recorded at an age of 28 days by mercury intrusion porosimetry under pressures of 0 - 200 MPa, and by a light optical microscope, respectively. The inside of the steel molds to test the properties of hardened concrete were lined with stiff vinyl to prevent interaction with the mold release oil. Immediately after casting, most specimens were then sealed using a plastic bag to prevent evaporation and then cured at room temperature.

3. Test Results and Discussion

Test results are summarized in **Table 1**. As the foam content generally plays a leading factor in the foamed concrete with excessively low density [12], precisely controlling the designed foam volume being pumped into the cementitious slurry is essential to achieve the targeted properties of such concrete. However, an error between the actual foam content and designed value would be sometimes occurred because the density of pre-formed foam somewhat varies with the aerating time, pressure of compressed air connected with the foam generator and atmospheric temperature. In the present test, the differences between the actual and the designed foam volumes were no more than 3%, indicating that the mixing and pumping of the pre-formed foam were successfully achieved as intended. Hence, the designed foam volume is used for the following discussion.

3.1. Flow of Fresh Concrete

All mixes showed the initial flow more than 180 mm, which is the minimum requirement value specified in KS F 4039 [11] (see **Table 3**). The value of initial flow of concrete mixes of Group I exceeded the diameter of table for measuring flow, though higher flow was visually observed with the increase in W/B ratio. The substitution of FA slightly contributed to improving the flow of AA GGBS foamed concrete, showing a higher flow with the increase in the substitution level (R_{FA}) of FA. The flow of fresh concrete demonstrates that the alkali activators selected for the present tests were effective in achieving high workability and preventing quick setting of the foamed concrete.

3.2. Defoamed Depth

The defoam in mixes of Group I almost did not occur, indicating that these specimens meet the requirement for Grade 0.6 of KS F 4039 [11]. On the contrary, $Ca(OH)_2$ and Na_2SiO_3-activated concrete mixes of Groups II showed relatively high defoamed depth. In particular, the defoamed depth increased with R_{FA}, indicating that the defoamed depth exceeded the minimum requirements for Grade 0.4 of KS F 4039 when the R_{FA} is above 15%. Severe defoam causes the deterioration of the thermal insulation capacity of the foamed concrete and cracking and settlement of the finishing lath mortar. From the comparison of activators used in Groups I and II, it can be inferred that Na_2SiO_3 as an activator is unfavorable to produce AA GGBS foamed concrete in terms of a burst of bubbles. The substitution of FA also accelerates the burst of bubbles in AA GGBS foamed concrete.

3.3. Dry Density

At the same unit binder content, the dry density of the foamed concrete increased slightly with the W/B ratio owing to the decrease in foam content, as given in **Table 1**. The effect of replacement of FA on the dry density of AA GGBS foamed concrete was negligible, although the specific gravity of FA is lower than that of GGBS. The concrete mixes of Group I commonly met the density requirements of Grade 0.4 specified in KS F 4039, whereas those of Group II was relevant to Grade 0.5. Yang et al. [10] proposed that the dry density of the foamed concrete is proportional to its nominal unit weight ($W_n = W_B + W_W + W_f$) of the plastic mix based on absolute volume, where W_B, W_W, and W_f are the weights per unit volume of the binder, water and pre-foamed foam, respectively. The present tests also confirmed that the dry density of the AA GGBS foamed concrete can be expressed in terms of its nominal unit weight, as shown in **Figure 2**.

3.4. Compressive Strength

All concrete mixes achieved the minimum strength requirements of Grade 0.5 specified in KS F 4039. As might be expected, the compressive strength of the AA GGBS foamed concrete decreased with the increase in the W/B ratio, as given in **Table 1**. In general, concrete mixes in Group II developed lower strength than those in Group I. This indicates that 10% $Ca(OH)_2$ and 4% $Mg(NO_3)_2$ activators is more favorable than 2.5% $Ca(OH)_2$ and 6.5% Na_2SiO_3 activators in developing the compressive strength of AA GGBS concrete. The substitution of FA also significantly influenced the compressive strength of AA GGBS foamed concrete, showing that the compressive strength increased up to R_{FA} of 15%, beyond which it turned to decrease.

3.5. Porosity and Pore Structure

Figure 3 shows the effect of the W/B ratio on the pore size distribution of AA GGBS foamed concrete. It was fail to measure the pore size distribution in mixes of Group II. The air-void structure of mixes tested is also shown in **Figure 4**. The macro capillaries ($50 \leq \phi < 50$ μm) and artificial air pores (50 μm $\leq \phi$) result from the deliberately entrained air and insufficient compaction,

Table 3. Quality and grade of foamed concrete for thermal insulation specified in KS.

Grade	Fresh concrete			Hardened concrete			
	Wet density of slurry (kg/m³)	Flow (mm)	Defoamed depth (mm)	Dry density (kg/m³)	Compressive strength (MPa)		Thermal conductivity (W/mK)
					7 days	28 days	
0.4	≥390		≤15	300 - 400	≥0.5	≥0.8	≤0.13
0.5	≥520	≥180	≤10	400 - 500	≥0.9	≥1.4	≤016
0.6	≥720		≤6	500 - 700	≥1.5	≥2.0	≤0.19

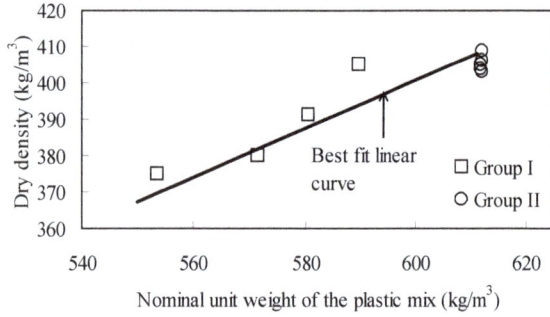

Figure 2. Relationship of nominal unit weight and dry density.

Designation	W/B ration (%)			
	40	45	47.5	50
Gel pores: $\phi < 10$ nm (%)	0.89	1.03	0.97	1.54
Micro capillaries: $10 \le \phi < 50$ nm (%)	14.85	12.99	11.51	10.47
Micro capillaries: 50 nm $\le \phi < 50$ μm (%)	32.05	34.54	36.21	41.56
Artificial air pores: 50 nm $\le \phi$ (%)	7.45	8.17	8.84	9.91
Total porosity (%)	55.24	56.72	57.53	63.48

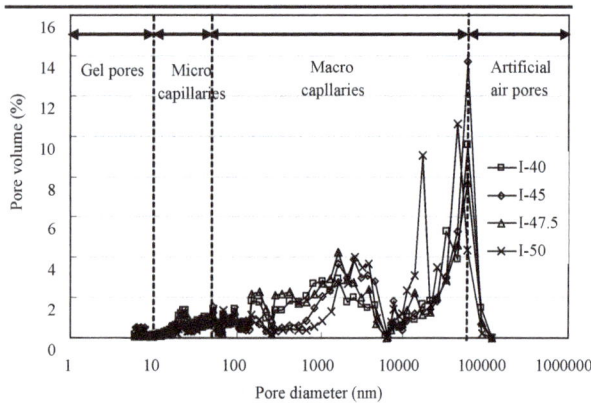

Figure 3. Effect of W/B ratio on the pore size distribution.

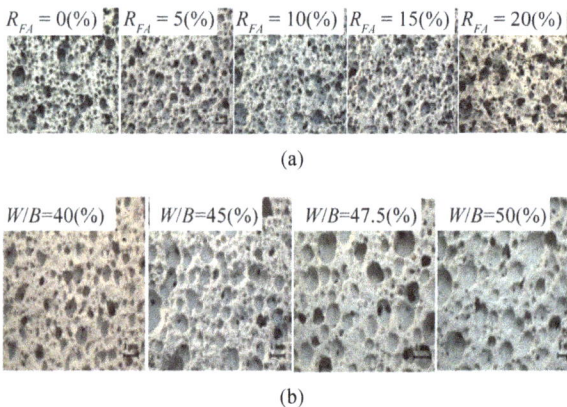

(a)

(b)

Figure 4. Air-void structure of AA GGBS foamed concrete. (a) Effect of W/B ratio; (b) Effect of R_{FA}.

where ϕ is the pore diameter. With the increase in the W/B ratio, the amount of micro capillaries decreased whereas those of macro capillaries and artificial air pores increased, which resulted in the increase in porosity. In addition, a few bigger-sized pores were found, and their number also increased in concrete with higher W/B ratio. On the other hand, the amount of bigger-sized pores decreased with the increase in R_{FA} up to 10%, beyond which it increased. The effect of test parameters on pore size distribution and air-void structure of the foamed concrete was coincident with the observation in compressive strength. This indicates that the increase of macro capillaries and artificial air pores are responsible for reduction in strength.

4. Environmental Impact

The environmental loads of foamed concrete were calculated in accordance with the lifecycle assessment (LCA) procedure specified in ISO 14040 series [15] based on the Korean lifecycle inventory (LCI) database [16]. The studied boundary condition was from the cradle to the pre-construction system including various contributions that are subdivided into the constituent phase, production phase using a mixer in the construction site, and transportation phase from the gate of the raw material-producing facility to the site. The LCI for a building material provides a collective data set that covers everything from the cradle to the grave. The CO_2 inventory for the concrete production phase was obtained from the conversion of energy sources consumed in the mixer. The procedure and typical examples of LCA for various environmental loads are explained in detail in literature [17].

Table 4 summarizes the magnitude of the environmental loads determined from the mixing proportion of each specimen based on the LCA procedure. The environmental load inventories obtained from the typical OPC foamed concrete used for cost comparisons are also given for comparisons. The W/B ratio, unit binder content and designed foam volume ratio of OPC foamed concrete were typically assumed to be 50%, 425 kg/m³, and 65%, respectively, based on the case investigation result [1]. Because the CO_2 inventory of Na_2SiO_3 are considerably higher than those of $Ca(OH)_2$ and $Mg(NO_3)$, the CO_2 emission of concrete mixes of Group II was commonly higher than those of mixes of Group I. The magnitude of the environmental loads of the AA GGBS foamed concrete tested is independent of the W/B ratio and R_{FA}.

The magnitude of the environmental loads of the foamed concrete tended to be remained constant regardless of the W/B ratio and R_{FA}. This is because the unit binder content, which governs the environmental load of concrete, was fixed in each group. Among the environmental load measures in each AA GGBS mix, the largest was the CO_2 emission which was followed by the consumption of natural gas. The CO_2 emission of AA GGBS

Table 4. Environmental load inventories of each concrete mix calculated from LCA procedure.

Specimen	Environmental load inventories (kg/m^3)								
	Emissions					Primary energy use			
	CO$_2$	CO	SO$_X$	NO$_X$	NH$_3$	Anthracite coal	Bituminous coal	Natural gas	Crude oil
I-40	3.18E+01	2.20E−02	3.45E−02	2.29E−01	3.62E−03	7.36E−01	4.04E−03	1.51E+01	2.44E+00
I-45	3.18E+01	2.20E−02	3.45E−02	2.29E−01	3.62E−03	7.36E−01	4.04E−03	1.51E+01	2.44E+00
I-47	3.18E+01	2.20E−02	3.45E−02	2.29E−01	3.62E−03	7.36E−01	4.04E−03	1.51E+01	2.44E+00
I-50	3.18E+01	2.20E−02	3.45E−02	2.29E−01	3.62E−03	7.36E−01	4.04E−03	1.51E+01	2.44E+00
II-F00	5.46E+01	2.97E−02	4.96E−02	1.56E−01	2.67E−03	2.10E−01	2.65E−03	8.91E+00	6.30E+00
II-F05	5.44E+01	2.90E−02	4.95E−02	1.55E−01	2.67E−03	2.10E−01	2.59E−03	8.91E+00	6.30E+00
II-F10	5.42E+01	2.83E−02	4.93E−02	1.54E−01	2.67E−03	2.10E−01	2.52E−03	8.91E+00	6.30E+00
II-F15	5.40E+01	2.76E−02	4.91E−02	1.54E−01	2.67E−03	2.10E−01	2.46E−03	8.91E+00	6.30E+00
II-F20	5.38E+01	2.68E−02	4.89E−02	1.53E−01	2.67E−03	2.10E−01	2.39E−03	8.91E+00	6.30E+00
OPC	4.43E+02	3.80E+01	2.54E−01	9.73E−01	6.29E−02	3.23E+00	7.66E+01	3.46E+00	1.35E+01

foamed concrete was evaluated to be considerably reduced by 93% for mixes of Group I, and 88% for mixes of Group II, compared with that of OPC foamed concrete. The reduction percentage of each environmental impact profile of AA GGBS foamed concrete relative to the typical OPC foamed concrete is plotted in **Figure 5**. The selected environmental impact category included abiotic depletion, global warming potential, acidification potential, eutrophication potential, photochemical oxidation potential, and human toxicity potential. The largest reduction percentage was found in the photochemical oxidation potential, being more than 99%. The reduction percentage was 87% - 93% for the global warming potential, 81% - 84% for abiotic depletion, 79% - 84% for acidification potential, 77% - 85% for eutrophication potential, and 73% - 83% for human toxicity potential. Overall, it can be concluded that the AA GGBS foamed concrete is promise as a sustainable building material with considerably reduced environmental impact.

5. Concluding Remarks

From the present experimental investigation and assessment of environmental impact, the following conclusions may be drawn:

1) The substitution of fly ash slightly contributed to improving the flow of AA GGBS foamed concrete.

2) Na$_2$SiO$_3$ as an activator is unfavorable to produce AA GGBS foamed concrete in terms of a burst of bubbles. The substitution of FA also accelerates the burst of bubbles in AA GGBS foamed concrete.

3) The dry density of the foamed concrete increased slightly with the W/B ratio, whereas the effect of replace-

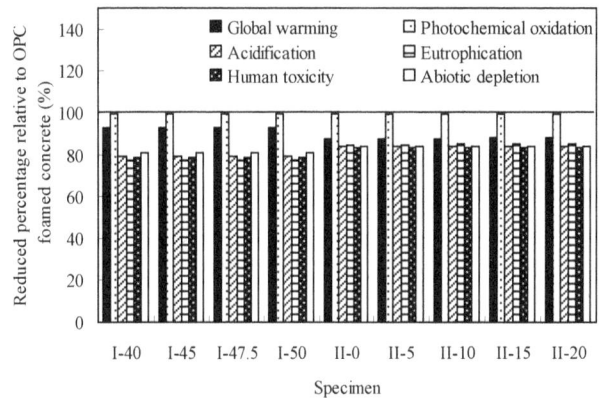

Figure 5. Reduction of each environmental impact profile of AA GGBS concrete relative to the typical OPC foamed concrete.

ment of fly ash on the dry density was negligible.

4) The compressive strength of the AA GGBS foamed concrete increased with the increase in R_{FA} up to 15%, beyond which it decreased.

5) With the increase in the W/B ratio, the amount of micro capillaries decreased whereas those of macro capillaries and artificial air pores increased, which resulted in the increase in porosity.

6) The magnitude of the environmental loads of the AA GGBS foamed concrete tested is independent of the W/B ratio and R_{FA}. The largest reduction percentage was found in the photochemical oxidation potential, being more than 99%. The reduction percentage was 87% - 93% for the global warming potential, 81% - 84% for abiotic depletion, 79% - 84% for acidification potential, 77% - 85% for eutrophication potential, and 73% - 83% for human toxicity potential.

6. Acknowledgements

This work was supported by the Basic Science Research Program through the National Research Foundation of Korea (NRF: 2009-0067189). The sponsor and technical participation of Hyundai Amco Co during the preparation of the test is greatly appreciated.

REFERENCES

[1] J.-K. Song and K.-H. Yang, "Development of Environmental-Friendly High-Performance Floor System," Technical Report, Department of Architectural Engineering, Chonnam National University, Gwangju, 2012.

[2] E. Gartner, "Industrially Interesting Approaches to 'Low-CO_2' Cements," *Cement and Concrete Research*, Vol. 34, No. 9, 2004, pp. 1489-1498.

[3] B. L. Damineli, F. M. Kemeid, P. S. Aguiar and V. M. John, "Measuring the Eco-Efficiency of Cement Use," *Cement & Concrete Composite*, Vol. 32, No. 8, 2010, pp. 555-562.

[4] F. Pacheco-Torgal, J. Castro-Gomes and S. Jalali, "Alkali-Activated Binders: A Review-Part 1. Historical Background, Terminology, Reaction Mechanism and Hydration Products," *Construction and Building Materials*, Vol. 22, No. 7, 2008, pp. 1305-1314.

[5] P. Duxson, A. Fernández-Jiménez, J. L. Provis, G. C. Lukey, A. Palomo and J. S. J. van Deventer, "Geopolmer Technology: The Current State of the Art," *Journal of Material Science*, Vol. 42, No. 9, 2007, pp. 2917-2933.

[6] C. Shi, P. V. KrddShi and D. Roy, "Alkali-Activated Cements and Concretes," Taylor and Francis, London, 2006.

[7] S. D. Wang, X. C. Pu, K. L. Scrivener and P. L. Pratt, "Alkali-Activated Slag Cement and Concrete: A Review of Properties and Problems," *Advanced Cement Research*, Vol. 7, No. 27, 1995, pp. 93-102.

[8] J. Davidovits, "Geopolymer: Chemistry & Applications," Géopolymère, 2008.

[9] H. Esmaily and H. Nuranian, "Non-Autoclaved High Strength Cellular Concrete from Alkali Activated Slag," *Construction and Building Materials*, Vol. 26, No. 1, 2012, pp. 200-206.

[10] K.-H. Yang, K.-H. Lee, J.-K. Song and M.-H. Gong, "Development of Alkali-Activated Slag Foamed Concrete for Thermal Insulation," *Cement & Concrete Composite*, Submitted for Publication, 2013.

[11] KS F 2459, F 4039, "Korean Industrial Standard: Testing Concrete," Korean Standards Information Center (KS), Seoul (in Korean), 2006.

[12] K. Ramamurthy, E. K. K. Nambiar and G. I. S. Ranjani, "A Classification of Studies on Properties of Foam Concrete," *Cement & Concrete Composite*, Vol. 31, No. 6, 2009, pp. 388-396.

[13] ASTM C796-97, "Annual Book of ASTM Standards: V. 4.02," ASTM International, 2012.

[14] D.-H. Lee, M.-H. Jun and J.-S. Ko, "Physical Properties and Quality Control of Foamed Concrete with Fly Ash for Cast-in-Site," *Journal of Korea Concrete Institute*, Vol. 13, No. 1, 2001, pp. 69-76 (in Korean).

[15] ISO 14040, "Environmental Management-Life Cycle Assessment—Principles and Framework," International Standardisation Organisation 2006.

[16] Korea LCI Database Information Network, (in Korean). http://www.edp.or.kr/lcidb

[17] K.-H. Yang, J.-K. Song and K.-I. Song, "Assessment of CO_2 Reduction of Alkali-Activated Concrete," *Journal of Cleaner Production*, Vol. 39, No. 1, 2013, pp. 265-272.

3

Crime Prevention in Ethnic Areas Focusing on Crime Prevention through Environmental Design

Seok-Jin Kang

School of Architecture, Gyeongsang National University, Jinju-Si, Korea.

ABSTRACT

The purpose of this study is to consider crime prevention measures in ethnic areas focusing on Crime Prevention through Environmental Design (CPTED) by an analysis of crime data and field survey. In this study, it was found that the main type of foreign crime that occurred in the research area was violence, and crimes committed by Koreans, which were mainly violence and crimes such as burglary, theft, robbery, and sexual offences, occurred steadily. Because it was found that crimes were related to the urban planning elements comprised of land use such as traditional market, inn, pub, and complicated space structure and the architectural design for natural surveillance and security facilities such as CCTV, lighting, alarm, and target hardening device, a new strategy for crime prevention design should include street environmental management, improvement of commercial facilities, and reinforcement security device of each buildings has to be spread through support of policy. In conclusion it was thought that CPTED would be a valuable measure to prevention crime and support community activities in ethnic area as expecting an improvement of physical environment and resident participatory for safer community.

Keywords: Crime Prevention through Environmental Design (CPTED); Foreign Crime; Ethnic Area; Hot-Spot; Architectural Design; CCTV

1. Introduction

One of the most significant current discussions in Korea is the social problem of the proliferation of the number of foreign workers' residences [1,2]. More specifically, the foreign residential areas in Korea have especially designated, and the discussion is centered on whether crime in those areas is related to their personal and socio-cultural characteristics or physical environmental components [1-5]. In the field of architecture and urban planning, Crime Prevention through Environmental Design (CPTED) is an important research theme that studies environmental measures consisting of crime-inducing or deterring factors [6]. However, so far, there has been little discussion about the crime of foreign residents within CPTED.

In this respect, it is necessary to analyze environmental factors related to crime in the ethnic areas. With this problem in mind, this study explores crime prevention measures through a theoretical study of CPTED, a literature review, an analysis of crime data, a field survey of ethnic areas, and a comprehensive analysis of the relationship between criminal acts and environmental factors.

The scope and process of this study are as follows: 1) a statistical and spatial analysis of the National Police Agency's crime data; 2) a literature review and composition of a checklist for the field survey; 3) a field survey based on the analysis of the crime data; and 4) discussion of crime prevention measures focusing on CPTED.

2. Theoretical Study and Literature Review

2.1. CPTED

CPTED has emerged and been given considerable academic and administrative attention recently as a new paradigm and an important dimension in crime prevention. CPTED, as proposed by C. Ray Jeffery (1972) [7], is a theory for crime prevention and community activation composed of five design principles: natural surveillance, access control, territoriality reinforcement, activity support, and maintenance [6]. CPTED is applied in the area of architectural and urban planning to eliminate of criminal opportunities through a comprehensive analysis of three main elements that lead to crime: motivated criminals, vulnerable victims, and environmental opportunities [8-10]. CPTED is included in the field of Environmental Criminology because it is theory about the relationship between the environment and crime.

2.2. Literature Review

(1) Studies on the CPTED

Studies on the CPTED originating in the U.S. and Great Britain have increasingly put stress on the development of architectural and urban design guidelines, an analysis of the environmental characteristics of crime hot-spots, and the crime prevention effect and industries related to CPTED, which are institutionally supported in developed countries. According to the foreign countries' studies, there are specific environmental factors that affect offenders' behavior and decision making. These factors are directly or indirectly related to psychological and behavioral elements, such as the residents' attachment to their homes, neighborhood relationship, informal social control, and territoriality (e.g. sense of ownership) and to physical elements, such as estate layout and dwelling characteristics [11-22]. However, studies on the relationship between crime and environmental elements in ethnic areas that focus on CPTED or environmental criminalogy are considered insufficient.

Conversely CPTED in Korea has been developing since the 2000s, and most of the studies are focused on crime and disorder in apartment complexes [23]. Studies on the crime pattern analysis of and CPTED application to detached houses, multi-family houses, and residential areas are unsatisfactory. Similar to the results of foreign countries' studies, Korean studies have verified that crime and fear of crime are related to factors including physical elements, such as the ratio of dwellings to other buildings, estate layout (e.g., the location of windows and entrances for natural surveillance), dwelling characteristics, alleyway structures, the number of intersections, and the density of street lighting and CCTV, and to psychological and behavioral elements, such as territoryality (range of living conditions), the solidarity of the community, and natural or organized surveillance [23, 24].

(2) Studies on ethnic areas

Studies on foreign crime in ethnic areas [1-5] can be organized into the sub-categories of "the study on the relation between ethnic places and dwelling environment" and "the study on the public security and foreign crime in ethnic places." The former sub-category has mainly been studied in the fields of geography, architecture, and urban planning, whereas the latter sub-category has been studied in criminology, penology, and police public administration. Although interdisciplinary studies about the characteristics of foreign crime and ethnic areas focusing on CPTED have not been sufficient, the literature review about them can be summarized as follows: 1) Foreigners are clustered in the specific areas; 2) Foreign residential areas in Korea are deteriorating due to a lack of urban infrastructure and the location of clustered multi-family housings adjacent to the manufacturing industries and traditional markets; 3) The rate of foreign crimes has increased steadily; 4) The causes of foreign crimes seem to be linked to social and physical factors such as cultural conflict, racial discrimination, and the vulnerability of the environment.

3. Summary of This Research

3.1. Method of Field Survey and Data Analysis

(1) Method of field survey

Principles of the research area selection as follows: 1) Foreigners are clustered in the specific areas; 2) There is a similarity between ethnic areas in respect of size and population composition; 3) They are different in physical characteristics of environment between ethnic areas. To select the research areas with these conditions, reported foreign crimes were reviewed with the help of the National Police Agency. The researcher focused on ethnic areas in which foreign migrant workers were clustered within city limits and selected Garibong-dong in Seoul and Wongok-dong in An-san. There are many migrant workers from China in Garibong-dong and foreign migrant workers from Southeast Asia in Wongok-dong. These areas have attracted growing interest in their various crime problems and have experienced a remarkable increase in the number of foreign workers since the early 1990s. Field survey was carried out in target areas by checklist in October 2012 and the checklist was made from theoretical-based on CPTED principles.

(2) Method of data analysis

In this study, the general tendency of crime in the years 2001-2011 was analyzed, and crime hot-spots of the five types of crime (murder, burglary, rape, robbery, and violence) in the research areas in 2011 were analyzed. The Geographic Information System (GIS) ArcMap 9.3 was used to analyze the crime hot-spot and the Statistical Package for Social Science (SPSS) was used to analyze the crime frequency.

3.2. The Trend of Foreign Crime in Korea

According to the National Police Agency's statistical data on foreign crimes, the rate of foreign crimes increased remarkably between 2001 (4328 arrests) and 2011 (27,144 arrests) (refer to **Figure 1**). The national origins of foreign offenders in 2011 were ranked as follows: China (15,667 persons, 58%); Vietnam (2438 persons, 9.1%); the United States (1788 persons, 6.6%); Mongolia (1,503 persons, 5.6%); and Thailand (944 persons, 3.5%). The incidence of the five types of crime (murder, burglary, rape, robbery, and violence) committed by foreign offenders among total crimes was higher than that of domestic offenders. The rate of the five types

person

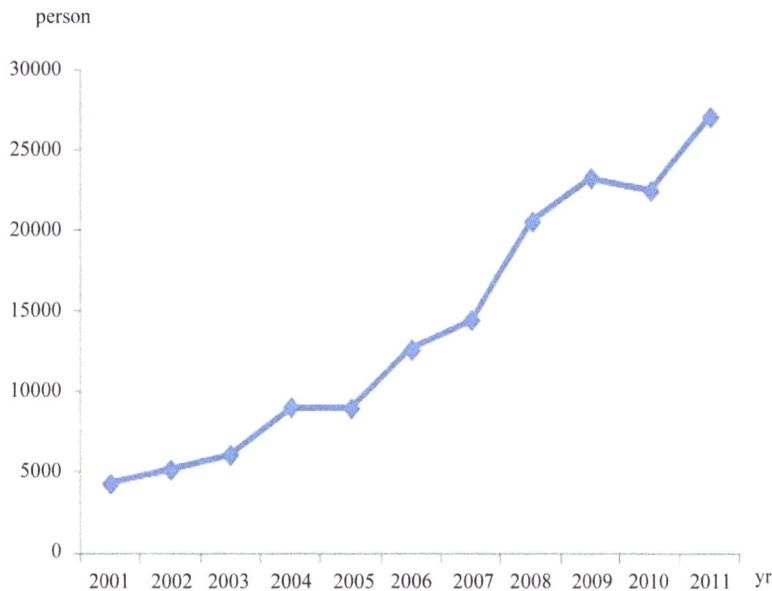

Figure 1. Trend of foreign crimes in Korea in the years 2001-2011.

of crimes by foreign offenders in 2011 was 37.8%, whereas the rate by domestic offenders was 28.0%. Of the five types of crime, the rate of violence was the highest.

4. Discussion of Results

The hot-spots were inferred through the spatial analysis of foreign residents, and the field survey was carried out according to the results of the hot-spot analysis. The field survey checklist was composed of CPTED principles, and a Likert scale (a three-point scale) was used to assess comprehensive items. The field survey was carried out on the streets (or alleys), in residential areas, and in commercial areas through the research's subjective judgment.

4.1. Case of Garibong-Dong

(1) Analysis of crime spots

The number of crimes that occurred in 2011 was 342. The cases of Korean crime numbered 223 (burglary - 19, robbery and sexual offence—7, violence—197), and the case of foreign crime numbered 119(burglary, robbery, and rape—8, violence—111). Foreign crimes were linearly concentrated in the surroundings of commercial facilities and the traditional market located between a subway station and a five-way crossing. In addition, some crimes were occurred in a scheduled redevelopment area. The hot-spots of crimes committed by Korean were especially concentrated in the deteriorating residential areas as shown in **Figure 2**.

(2) Field survey results

The present conditions of alleyway, residential area,

and commercial area in Garibong-dong are as shown in **Figure 3**.

a) Alley and street

In the natural surveillance aspect, there were insufficient streets lighting and narrow and curved paths that connected continually and repeatedly. Moreover, the cars parked randomly on the street obstructed a pedestrian passage and sightline to the surroundings. In the access control aspect, there were insufficient CCTV and burglar alarms. In the territoriality reinforcement aspect, the information map and signboard were inadequate in terms of number, location, and legibility. In the activity support aspect, there were not leisure spaces and sport facilities for the local community. It is supposed that the narrow and complicated paths and spatial structure contributed to these conditions. In the maintenance aspect, there were insufficient cleaning of the paths and management of the street facilities.

b) Residential area: detached houses, multi-family houses, and row houses

Because deteriorated houses of various sizes stood close together in this neighborhood and the traditional market was adjacent to the houses, there was a considerable transient population and, as a results, the possibility for various crimes to occur on the condition of anonymity. According to the field survey results, there were many problems involving street lightings, widow positioning, and piloti in the natural surveillance aspect. As for access control, there were problems related to main entrances and windows without security devices, deserted houses, and side walls with exposed gas pipes. In the maintenance aspect, it was necessary to repair the deteriorated houses.

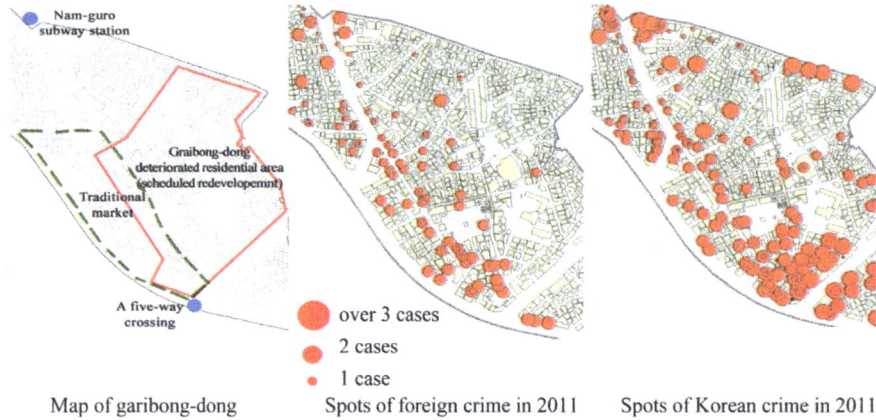

Map of garibong-dong   Spots of foreign crime in 2011   Spots of Korean crime in 2011

Figure 2. Crime spots in Garibong-dong in 2001.

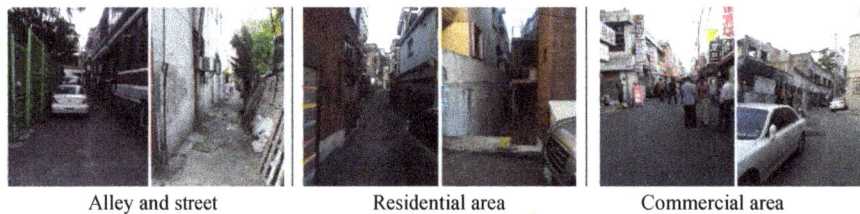

Alley and street   Residential area   Commercial area

Figure 3. The present conditions of Garibong-dong.

c) Commercial area

There were many problems concerning the location of lighting around the commercial facilities, windows partially obstructed by advertising media, and some commercial facilities' inadequate positioning, which created entrapment spots in the natural surveillance aspect.

In the access control aspect, there were problems with the insufficiency of security devices in each building and CCTV around the traditional market. In the maintenance aspect, it was necessary to repair the deteriorated buildings, messy streets, and signboards of the commercial buildings to create a safe milieu.

4.2. Case of Wongok-Dong

(1) Analysis of crime spots

The number of crimes that occurred in 2011 was 276. The cases of Korean crime numbered 126, and the case of foreign crime numbered 150 (robbery—8, sexual offences—6, violence—136). Foreign crimes occurred all over the target area, such as around inns, the traditional market, the neighborhood park, and the residential area. However, Korean crimes tended to be concentrated in a specific area as shown in **Figure 4**.

(2) Field survey results

There were many foreigners of various nationalities residing in Wongok-dong. The research area encompassed the main street and An-san subway station in the south, Wongok-park (neighborhood park) in the north, and the traditional market throughout the area. This area

has attracted foreigners due to political concerns and the availability of multi-family rental properties for foreigners in the low-income bracket. The present conditions of alleyway, residential area, and commercial area in Wongok-dong are as shown in **Figure 5**.

a) Alley and street.

Because the alley was wide and visible, there were many opportunities for natural surveillance. As for access control, there was CCTV in the intersection, but the number of CCTVs was insufficient for crime prevention. In the maintenance aspect, the conditions on the street and of all sorts of street facilities were favorable. However, territoriality reinforcement was insufficient for creating a small community space for residents to enjoy and for projecting public designs on the walls of buildings.

b) Residential area: detached houses, multi-family houses, and row houses

A vulnerable environment was fostered under building spaces and in the narrow spaces between buildings in the form of blind spots in the natural surveillance aspect. In the access control aspect, main entrances and windows were vulnerable because of a lack of security devices, and side walls were vulnerable because of exposed gas pipes. In the maintenance aspect, the residential area was well managed.

c) Commercial area

There were some problems with the lightings around commercial facilities and with windows partially obstructed by advertising media. As for access control, this

over 3 cases   2 cases   1 case

Map of wongok-dong   Spots of foreign crime in 2011   Spots of Korean crime in 2011

Figure 4. Crime spots in Wongok-dong in 2001.

Alley and street   Residential area   Commercial area

Figure 5. The present conditions of Wongok-dong.

area was favorably assessed in terms of a security kiosk and CCTV in the main intersection of the residential area and traditional market. In the maintenance aspect, the commercial area and traditional market were well managed.

4.3. Comprehensive Review of the Field Survey and Hot-Spot Analysis

A summary of the field survey and hot-spot analysis is as follows. The main type of foreign crime that occurred in the research area was violence, and the hot-spot was concentrated in the area surrounding the subway stations, traditional markets, merrymaking places, and residential areas adjacent to commercial facilities. There were many problems in Garibong-dong (refer to **Table 1**), but Wongok-dong was favorably assessed from the view of CPTED (refer to **Table 2**).

In the case of Garibong-dong, there were insufficient security facilities such as lighting and CCTV, and the poor condition of the residential area was attributed to the proliferation of narrow and curved paths, which connected repeatedly, and to the close proximity of deteriorating houses of houses of various sizes.

In the case of Wongok-dong, there were many foreigners of various nationalities and residential and commercial facilities for them. Also, there were visible and

well managed streets arranged in a lattice configuration and sufficient security facilities due to the fact that it was a planned city, as opposed to Garibong-dong. However, there were problems related to piloti space under the buildings, narrow spaces with blind spots between the buildings, main entrances and windows without security devices, and side walls with exposed gas pipes. According to these results, it was inferred that the occurrence of crime was related to land use, street characteristics, and security facilities, similar to the findings discussed in the literature review. Foreigners lived in the residential area adjacent to the traditional market, where commercial facilities stood close together. Therefore, it seemed that the environmental elements that cause crime existed in those areas.

In contrast, crimes committed by Koreans, which were mainly violence and crimes such as burglary, theft, robbery, and sexual offences, occurred steadily. It was inferred that the environment had influenced various crimes, because the spaces that were assessed to be vulnerable to crime by the field survey were similar to the hot-spots for crime committed by Koreans.

4.4. Crime Prevention Measures for Foreign Residential Areas

The following measures are deemed necessary for pre-

Table 1. Crime risk assessment of Garibong-dong by field survey.

CPTED principles			Checklist	Assessment		
				inadequate	average	adequate
Alley and street	Natural surveillance	Street lighting	The luminous intensity of a street lighting is optimized.	○		
			The location of a street lighting is optimized.	○		
			The interval between street lighting and other street facilities is proper for surveillance.		○	
		Space structure	A narrow alley is not connected continually and repeatedly.	○		
			A curved alley is not connected continually and repeatedly.	○		
			There is a safe mioor on the curved or cross alley.	○		
			A parking lot on the alley is located on the proper area for surveillance.		○	
	Access control	Crime prevention facility	CCTV is installed optimally on the vulnerable alley and area.	○		
			CCTV is installed on the gateway of district.	○		
			An alarm is installed on the vulnerable alley and area.	○		
			Design of an alarm is legible.	○		
		Other facility	A crime prevention post is located on the proper area for surveillance and access control.	○		
			The dead-end alley and entrapment spot are to be access control.	○		
	Territoriality reinforcement	Information facility	There is an information map or sighboard on the area.	○		
			Design of the information facility is legible.	○		
		Other facility	Waste pieces on the street ares used for community activity.	○		
			Environmental design with wall painting or architectural ornaments is applied on a disorder area for safe district milieu.	○		
	Activity support	Facility layout	There are activity support facilities on the area.	○		
			Activity support facilities are located on the proper area.	○		
	Maintenance	Cleaning and repair	Cleaning conditions on the alley, facility, and building are fine.	○		
			Street facilities are well managed.	○		
Residential area	Natural surveillance	Wndow	It is installed properly for natural surveillance.		○	
		Entrance	There is a door with lighting.	○		
		Wall	It is installed properly for natural surveillance.	○		
	Access control	Entrance	There is the safe door equipped with security device.	○		
			The space between buildings is controlled against intrusion.	○		
		Side wall	Security lock on the window is fitted for crime prevention.	○		
			External pipes on the side wall are concealed.	○		
		Extra space	It is not allowed a person access to an abolished house.	○		
	Territoriality reinforcement	Informantion	An address signage on the house is legible.	○		
		Extra facility	Extra space around a house is used for territoriality.	○		
	Maintenance	Management	A house is well managed.	○		
Commercial area	Natural surveillance	Wndow and entrance	They are installed properly for natural surveillance.	○		
			There are lightings on the door and window.	○		
			Advertising materials on the door and window are removed for natural surveillance.	○		
	Access control	Entrance	CCTV is installed around a store.	○		
	Maintenance	Management	A shop is well managed.	○		

Table 2. Crime risk assessment of Wongok-dong by field survey.

CPTED principles			Checklist	Assessment		
				Inadequate	Average	Adequate
Alley and street	Natural surveillance	Street lighting	The luminous intensity of a street lighting is optimized.		○	
			The location of a street lighting is optimized.		○	
			The interval between street lighting and other street facilities is proper for surveillance.		○	
		Space structure	A narrow alley is not connected continually and repeatedly.			○
			A curved alley is not connected continually and repeatedly.			○
			There is a safe mioor on the curved or cross alley.	○		
			A parking lot on the alley is located on the proper area for surveillance.			○
	Access control	Crime prevention facility	CCTV is installed optimally on the vulnerable alley and area.		○	
			CCTV is installed on the gateway of district.		○	
			An alarm is installed on the vulnerable alley and area.	○		
			Design of an alarm is legible.	○		
		Other facility	A crime prevention post is located on the proper area for surveillance and access control.		○	
			The dead-end alley and entrapment spot are to be access control.		○	
	Territoriality reinforcement	Information facility	There is an information amp or sighboard on the area.		○	
			Design of information facility is legible.	○		
		Other facility	Waste pieces on the street are used for community activity.	○		
			Environmental design with wall painting or architectural ornaments is applied on a disorder area for safe district milieu.	○		
	Activity support	Facility layout	There are activity support facility and space on the area.		○	
			Activity support facilities are located on the proper area.		○	
	Maintenance	Cleaning and repair	Cleaning conditions on the alley, facility, and building are fine.			○
			Street facilities are well managed.			○
Residential area	Natural surveillance	Wndow	It is installed properly for natural surveillance.		○	
		Entrance	There is a door with lighting.	○		
		Wall	It is installed properly for natural surveillance.		○	
		Entrance	There is the safe door equipped with security device.		○	
	Access control	Side wall	The space between buildings is controlled against intrusion.	○		
			Security lock on the window is fitted for crime prevention.		○	
			External pipes on the side wall are concealed.	○		
		Extra space	It is not allowed a person access to an abolished house.			○
	Territoriality reinforcement	Informantion	An address signage on the house is legible.		○	
		Extra facility	Extra space around a house is used for territoriality.	○		
	Maintenance	Management	A house is well managed.			○
Commercial area	Natural surveillance	Wndow and entrance	They are installed properly for natural surveillance.		○	
			There are lightings on the door and window.	○		
			Advertising materials on the door and window are removed for natural surveillance.	○		
	Access control	Entrance	CCTV is installed around a store.		○	
	Maintenance	Management	A shop is well managed.		○	

vention of crime in foreign residential areas. Comprehensive environmental improvement is necessary for crime prevention in Garibong-dong, where there are insufficient urban infrastructures and deteriorating residential facilities. CCTV, public lighting, and security alarms especially needs to be increased and the traditional markets need to be modernized.

Because it will take a certain period of time to improve the environment of the deteriorating district, it is necessary to make a community space for activity support and to reinforce patrol by the police or self-defense watchmen in the short term, and to guide the architectural design of the area according to CPTED principles and to convert an inn and a pub in the residential area into wholesome land use in the long term.

To put it concretely, the target hardening measures must be applied to the narrow spaces with blind spots between the buildings, the main entrances, and windows without security devices, and side walls with exposed gas pipes to prevent crimes such as burglary, theft, and robbery, in consideration of the deteriorating buildings and space structure.

In the case of Wongok-dong, it is necessary to reinforce crime prevention measures in the commercial area verified as a hot-spot, to close-in the residential area and to increase the number of lighting and alarm and to improve degree of visibility on the commercial facilities' window in view of CPTED.

And likewise the Garibong-dong case, it is necessary to apply concept of the target hardening and natural surveillance measures on a blind spot between the buildings, piloti space, and side wall space against burglary and robbery.

5. Conclusions

Until now, the crime prevention strategies proposed by government and municipality are summarized as CCTV installation and reinforcement of police patrol. However these strategies are not enough to deal with various crimes committed in everywhere. Under these circumstances, CPTED could be an alternative strategy for crime prevention to environment condition.

As CPTED application is important for establishing safety infra n society through environmental design and management [6,7,19], it is suggested to be considered a mandatory standard on architectural and urban planning for crime prevention.

In this study, much of crime related violence had been committed in the ethnic area. Moreover, a number of certain crimes such as burglary and robbery increased steadily. Because crimes are related to the urban planning elements comprised of land use such as traditional market, inn, pub, and complicated space structure and the

architectural design for natural surveillance and security facilities such as CCTV, lighting, alarm, and target hardening device, a new strategy for crime prevention design should include street environmental management, improvement of commercial facilities, and reinforcement security device of each buildings has to be spread through support of policy. Also the design for community activation is recommended to be applied on public space and installed CCTV and alarm with a siren on the gateway of district, an intersection, and entrapment spots.

Finally it is necessary to reinforce the function of existing self-defense watchmen and a night patrol by the police on the commercial area for foreigners. In conclusion it was thought that CPTED would be a valuable measure to prevention crime and support community activities in ethnic area as expecting an improvement of physical environment and resident participatory for safer community.

REFERENCES

[1]   C.-H. Kim and H.-K. Ahn, "Spatial Distribution and Causes of Foreign Residential Areas in Seoul Metropolitan Area by Immigration Circuits: focused on Professional and Working groups," *Journal of Korea Planners Association*, Vol. 46, No. 5, 2011, pp. 233-248.

[2]   S.-H. Park and S.-Y. Jung, "Spatial Distribution of Foreign Population and Policy Implication in South Korea," *The Korea Spatial Planning Review*, Vol. 64, No. 4, 2010, pp. 59-76.

[3]   S.-H. Park, Y.-A Lee, E.-R. Kim and S.-Y. Jung, "Reinventing Urban Policy in Response to Ethnic Diversity: A Report on Emerging Ethnic Places in South Korea," Korea Research Institute For Human Settlements, Anyang-Si, 2009.

[4]   J.-E. Jung, G.-S. Ha and J.-M. Jun, "Analysis on the Determinants of Residential Location Choice for Foreign Residents in the Seoul Metropolitan Region," *Journal of Korea Planners Association*, Vol. 46, No. 6, 2011, pp. 117-130.

[5]   J.-O. Kim, "A Study on the Felony of Foreign Laborer," *Korean Journal of Public Safety and Criminal Justice*, Vol. 37, No. 4, 2009, pp. 97-127.

[6]   T. D. Crowe, "Crime Prevention through Environmental Design: Applications of Architectural Design and Space Management Concepts," Butterworth-Heinemann, Oxford, 2000.

[7]   C. R. Jeffery, "Crime Prevention through Environmental Design," Sage Publications, 1972.

[8]   E. R. Groff, "Exploring the Geography of Routine Activity Theory: A Spatio-Temporal Test Using Street Robbery," Ph.D. Thesis, University of Maryland, College Park, 2006.

[9]   R. V. Clarke and M. Felson, "Routine Activity and Rational Choice," New Brunswick, New Jersey, 2004.

[10] R. V. Clarke, "Situational Crime Prevention: Successful

Case Studies," 2nd Edition, Criminal Justice Press, New York, 1997.

[11] B. B. Brown, "Territoriality, Street Form, and Residential Burglary: Social and Environmental Analyses," Ph.D. Thesis, University of Utah, Salt Lake City, 1983.

[12] B. Yuen, "Safety and Dwelling in Singapore," *Cities*, Vol. 21, No. 1, 2004, pp. 19-28.

[13] P. L. Brantingham and P. J. Brantingham, "Nodes, Paths and Edges: Considerations on the Complexity of Crime and the Physical Environment," *Journal of Environmental Psychology*, Vol. 13, No. 1, 1993, pp. 3-28.

[14] E. R. Groff and N. G. La Vigne, "Mapping an Opportunity Surface of Residential Burglary," *Journal of Research in Crime and Delinquency*, Vol. 38, No. 3, 2001, pp. 257-278.

[15] J. E. Macdonald and R. Gifford, "Territoriality Cues and Defensible Space Theory: The Burglar's Point of View," *Journal of Environmental Psychology*, Vol. 9, No. 3, 1989, pp. 193-205.

[16] K.-H. Lee, "Community and Burglary in the Urban Residential Street Block: An Environmental Analysis," Ph.D. Thesis, University of Wisconsin-Milwaukee, 1992.

[17] L. Brunson, F. E. Kuo and W. C. Sullivan, "Resident Appropriation of Defensible Space in Public Housing," *Environment and Behavior*, Vol. 33, No. 5, 2001, pp. 626-652.

[18] M. Taylor and C. Nee, "The Role of Cues in Simulated Residential Burglary," *British Journal of Criminal*, Vol. 28, No. 3, 1988, pp. 396-401.

[19] O. Newman, "Defensible Space," Macmillan Publishing Co. Inc., London, 1972.

[20] R. B. Taylor and S. D. Gottfredson, "Territoriality, Defensible Space, Informal Social Control Mechanical, and Community Crime Prevention," Johns Hopkins University, Baltimore, 1978.

[21] R. B. Taylor, S. A. Shumaker and S. D. Gottfredson, "Neighborhood-Level Links between Physical Features and Local Sentiments," *Journal of Architectural and Planning Research*, Vol. 2, 1985, pp. 261-275.

[22] W. M. Rohe and R. J. Burby, "Fear of Crime in Public Housing," *Environment and Behavior*, Vol. 20, No. 6, 1988, pp. 700-720.

[23] S.-J. Kang and K.-H. Lee, "A Research on Creating Crime-Safe Environment Through Enforcing the Sense of Community in Urban Residential Area," *Journal of the Architectural Institute of KOREA*, Vol. 23, No. 7, 2007, pp. 97-106.

[24] S.-J. Kang and K.-H. Lee, "A Study on the Relationships of Outdoor Space Activation and the Experienced Crime Victimization Rate in Multi-Family Housings," *Journal of the Architectural Institute of KOREA*, Vol. 20, No. 2, 2004, pp. 71-78.

Journal of Building Construction and Planning Research: A New Platform for Building Researchers

Gwang-Hee Kim

Department of Plant & Architectural Engineering, Kyonggi University, Suwon-Si, South Korea.

Building construction projects include design, financial, estimating, environmental consideration, and legal review [1]. Building engineering is the application of theory, knowledge, technology, etc. to building construction. Building engineering can be classified into three categories: Structural Engineering, Mechanical, Electrical, and Plumbing (MEP), and Construction. Structural engineering includes the analysis and design of the structural frame. MEP engineering is a significant component of the building supply system and is the most important part of the building, which must be carefully coordinated before design and installation. In the field of architecture, construction consists of construction technology and construction management. Construction technology includes materials engineering, construction methods, etc. Also, construction management is a process encompassing the planning, coordination, and control of design, financial, and estimation of a project, from starting to completion.

Building planning is classified into three categories, which are architectural theory, architectural design, and sustainable architecture design. Architectural theory involves thinking, discussing and writing about architecture [2]. Architectural design is a plan for architecture, *i.e.* the written documentation and graphical descriptions of building projects. Sustainable architecture refers to eco-friendly design techniques in the field of architecture, and has recently been considered as a priority.

Recently, there has been quite a bit of multidisciplinary convergence between research fields. In building construction and planning, there is a tendency to focus research on the fields of construction engineering and architectural planning. But building construction and planning research clearly cannot be separated from each other, and as such, the cooperation between each field in practices of building construction and planning as well as in research is needed. Building construction and planning

is one of the composite arts. Research in the specific field of construction and planning research has been published in specific journals; that is, the scope of journals is quite narrow. Building construction and planning is quite a broad area of study, and as such it is not possible to cover all areas with research papers in one journal. However, the tendency to separate the various fields of building construction and planning may be undesirable for the progress of building construction, and does not match the recent trend of research.

Journal of Building Construction and Planning Research (JBCPR), a new open access journal, will include all areas of building construction and planning fields mentioned above, and will exclude original structural engineering and original MEP engineering. These fields are excluded because they are associated with buildings but are different from HVAC (heating, ventilation, and air conditioning), one of the factors affecting the building environment. All kinds of papers in all fields of building construction and planning, with the exception of these two fields mentioned above, are welcome to submit to JBCPR.

So what makes JBCPR significantly different from other civil engineering or building engineering journals? While other civil engineering or building engineering journals are concentrated on narrow, specific topics such as structural engineering, construction engineering, construction management, or environmental engineering, JBCPR is interested in the convergence of an associated field of study in building construction and planning. Therefore, JBCPR will be particularly interested in publishing articles on: 1) applications of information technology (IT) to building construction and planning; 2) green building design and construction for sustainable architecture; 3) crime prevention design through environmental design for community safety; 4) building design management for saving costs and shortening construction duration; 5) passive design for reducing CO_2 emission and energy consumption; 6) low energy embodied materials; 7) recycled materials; 8) healthy materials; 9) innovative construction methods; 10) lean construction management; 11) Human-centered construction management, etc. Nevertheless, JBCPR will also publish research papers that is not necessarily highly innovative or novel, and is also interested in: 1) technical notes; 2-5 page papers in which an author can either give an idea with a scientific basis but has not yet completed; 2) book reviews; 3) article reviews; 4) case studies of best practice, etc.

I wish to convey my sincere thanks to the Publisher for giving me the chance to launch this new journal. Also, I would like to thank all editorial members of JBCPR who have encouraged me in the launch of JBCPR, and who are serving as editorial board members of JBCPR. I hope that JBCPR will be a distinguished journal in the field of building construction and planning, and will become a main platform through which academic scholars, researchers, practitioners, and students can communicate with each other and share their knowledge, experience, study results, and information.

REFERENCES

[1]   Wikipedia, "Building Construction,"
      http://en.wikipedia.org/wiki/Construction

[2]   Wikipedia, "Architectural Theory,"
      http://en.wikipedia.org/wiki/Architectural_theory

Management Performance Evaluation Model of Korean Construction Firms

Donghoon Lee, Manki Kim, Sunkuk Kim[*]

Department of Architectural Engineering, Kyung Hee University, Seoul, Republic of Korea.

ABSTRACT

Corporate management performance evaluation currently focuses on financial aspects; however, it is necessary to identify and manage elements that contribute to increased economic values in the long run. When it comes to construction firms, most previous research did not cover weighting and estimation approaches for non-financial elements that ultimately influence financial status. In this research, the objective is to develop a management performance evaluation model for Korean construction firms. The model includes financial factors and non-financial factors. This research investigated actual data from Korean construction firms and classified their characteristics. This study is performed in two steps. First, this study derives KPIs for performance measurement techniques and weights the KPIs. And then, it applies the performance data of construction firms to the technique. The findings of this study show that Korean construction firms consider customers to be the foremost priority, converse to previous research which argued that the internal business process was the top priority. The performance measurement results can be fed back into strategies and plans to shed light on issues, reflect on management plans for subsequent years and modify mid to long-term strategies. Therefore, the developed model can help decision-makers effectively revise their management plans.

Keywords: Management Performance Evaluation; Key Performance Indicators; Balanced Scorecard; AHP; Fuzzy

1. Introduction

As Peter Drucker asserted, the adage "If you cannot measure it, you cannot manage it" is true in all corporate management activities, from manufacturing to IT. Performance measurement is an important management process that can be utilized to identify the fulfillment of annual management plans and provide feedback for subsequent management plans [1]. In the construction industry, performance measurements are complicated, as it is difficult to predict the productivity of projects due to the uncertainty in construction sites and contracting individually [2].

Especially, since Korea's economy has changed rapidly due to the sharp rise of raw material prices such as crude oil, the rapid increase in overseas plant construction contracts from 2004, the Korea construction firms need to be established effectively their management strategies to ensure the competitiveness including various characteristics such as financial and non-financial aspects. In addition, it is necessary to systematically measure management performance in order to investigate whether or not the annual plans are successful.

Traditionally, the management performance was measured in terms of financial aspects, including the current net profit, the investment rate, and the return on equity (ROE) [3]. However, with the increased development of IT in the 1990s, studies on measurement factors and methods began to focus on intellectual resources [4-6]. Meanwhile, in Korea, management performance measurement models for construction firms were developed [6-8]. However, the proposed key performance indicators (KPIs) were insufficient for measuring the non-financial management performances of Korean construction firms [9]. Previous studies did not suggest a systematic scoring method for non-financial factors. Therefore, to effectively measure management performance, it is essential to develop a performance measurement model that reflects the characteristics of Korean construction firms. In this study, our objective was to develop a management performance evaluation model (MAPEC) for Korean construction firms.

2. Previous Studies

Currently, most corporations focus on financial aspects

in their management performance measurements, as these are easy to measure [10]. However, it is difficult to evaluate the long-term potential of a corporation only on the basis of financial aspects. This potential indicates the non-financial aspects such as brand, image, work environment, etc. In other words, although these aspects do not reflect directly the financial factors such as profitability and growth, the aspects influences continuously on the management performance in the long term. Therefore, it is necessary to identify and manage qualitative factors that result in long-term economic value.

Beginning in the 1990s, with the advent of the concept of intellectual resources, many researchers investigated qualitative factors related to management performance measurement. As shown in **Table 1**, performance measurement indicators of construction project of [11] can be divided into four indicators: project efficiency, impact on customers, business success and preparation for the future. First, in the project efficiency phase, it is measured whether or not the project is completed on time and budget. Second, the satisfaction of customers is measured. Third, in the business success phase, it is measured whether or not the management performance is improved after completion of the project. Fourth, it is measured how reflect the performance to the management for the future.

C. S. Lim, and M. Z. Mohamed evaluated performances such as time, cost quality and safety from both micro and macro viewpoints of developers, contractors, and customers [12]. R. Atkinson measured performance by

dividing it into delivery and post-delivery phases [13]. First, in the delivery stage, the cost, time quality and efficiency is evaluated. Second, after completion of a project, the impact of performance to customers is evaluated. Although many studies have been conducted regarding management performance evaluation, only general indicators show in the studies, which did not consider the characteristics of Korea construction industry.

As an alternative to the traditional approach, [9] proposed a Balanced Scorecard (BSC), which included non-financial aspects related to performance evaluation in the long term. The BSC aims for a better understanding of the correct strategies and key performance factors for a more comprehensive insight into current businesses. As shown in **Table 2**, the BSC is divided into four perspectives (*i.e.*, financial, customer, internal business process and learning and growth). The BSC is appropriate for construction firms since the key performance indicators are defined in consideration of diverse environments such as market customers and culture. In this study, the BSC were applied to determine KPIs for Korean construction firms and then interviewed experts to estimate the weights of each KPI.

In terms of the performance measurement in construction industry, [14] suggested the construction performance measurement process conceptual framework. The framework represents the input and output of the process how to measure the performance as dividing into six perspectives such as financial, customer, internal business, innovation & learning, project, and suppliers. In addition, [15] analyzed the correlation between the change of managerial environment and the business performance of Korea construction firms. Although this study has been conducted regarding the management

Table 1. Previous literature of KPI [11-13].

Author	KPI	Criteria
Shenhar (1997)	Project efficiency	Short term measures completed on time? Within the specified budget?
	Impact on customer	Related to the customer and user Performance measures? Technical specifications
	Business success	Measures of time, quality and total improvement of organization performance
	Preparing for the future	Long term dimension Preparation organization Technological infrastructure
Lim and Mohamed (1999)	Micro viewpoint	Time, cost, quality, Performance, and safety
	Macro viewpoint	Time, satisfaction, utility, operation
Athinson (1999)	Delivery stage	Cost, time, quality, efficiency
	Post-delivery stage	Impact on customer Business success

Table 2. Four perspectives of BSC.

Classification	Perspective	KPI
Financial	How should we appear to our shareholders?	Revenue growth ROI Corporate earning Asset turnover Cost saving
Customer	How should we appear to our customers?	Customer return New customer acquisition Customer retention Customer satisfaction Market share
Internal business process	What business process must we excel at?	Product/service development New prototyping Customer management Business/work process Business environment
Learning & growth	How will we sustain our ability to change and improve?	Technology Knowledge sharing IT infrastructure Corporate culture

performance of Korea construction firms, it did not investigate how to measure the performance. [16] suggest a simulation method to show management effectiveness of various strategies.

3. Methodology

The MAPEC, intended to measure the management performance of Korean construction firms, consists of a hierarchical structure: Balanced scorecard indicator (BSCI) —Classified performance indicators (CPI)—Key performance indicator (KPI) (**Figure 1**). The collected corporate performance data are reflected in the hierarchical structure at the beginning of the MAPEC process. KPI analysis, which is positioned at the bottom of the hierarchical structure, assigns weights to the actual performance data and estimates the CPI. The CPI analysis then assigns estimated weights to produce a BSCI, and the comprehensive corporate management performance is estimated using the BSCI. Weights for different factors are estimated in a Fuzzy-Delphi Analytic Hierarchy Process (FD-AHP).

To achieve this study's objective, the following methodology was conducted: 1) CPIs and KPIs were obtained after reviewing previous studies and actual management measurement data; 2) CPIs and KPIs were selected after checking for duplication and omission; 3) according to the selected indicators, a management performance measurement hierarchy was proposed, and the weights of all of the indicators were estimated using an FD-AHP analysis; 4) each KPI score was evaluated by applying the scoring distribution methods proposed in this study; 5) two Korean construction firms were analyzed and evaluated using the MAPEC developed in this study.

It is difficult to develop an effective performance measurement model that encompasses all kinds of construction firms, such as those that design apartments, infrastructure, and plant development projects, as they cater to different customers and require different KPIs and CPIs for performance measurement [17]. Therefore, the scope of this study was limited to model development applicable to the top 30 corporations in Korea. Among these corporations, this study focused on firms with both concurrent overseas and domestic projects.

To analyse the experience of experts, Saaty's Analytic Hierarchy Process (AHP) method has been used widely. However, when the traditional AHP utilizes, the correlation between factors is not considered and the uncertainty and errors would have to determine ranks of factors. The problem such rank reverse may occur. The Fuzzy-Delphi AHP (FD-AHP) method is new decision-making model based on AHP method. In the decision-making issue, by using the pair-wise comparison method, the FD-AHP can determine the optimum alternative among various options [18]. This study utilizes FD-AHP method to establish management performance evaluation model.

4. MAPEC Development

4.1. CPI and KPI

In order to select the CPIs and KPIs, 10 experts were interviewed regarding the performance measurement factors from the top five Korean corporations. As shown in **Table 3**, among them, three experts are principals of the corporations and two experts are the head of management division. In addition, the rest of experts are the project managers having experience of over 10 years. After interviewing the experts, then the factors were classified and adjusted by the experts.

Based on the CPIs and KPIs, a survey was given to the top management executives in major Korean construction firms, asking for estimations of the weights of each indicator using pair-wise comparison. **Table 4** represents the BSCI, CPI, and KPI developed in this study. Finally, 12 CPIs and 31 KPIs were suggested as detailed indicators of the BSCI for Korean construction firms.

4.2. Weight Estimation of BSCI, CPI, and KPI

MAPEC consists of indicators, with weights assigned in

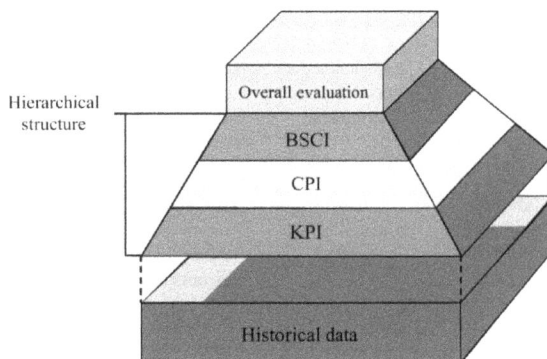

Figure 1. MAPEC concept.

Table 3. Summary of the respondents' demographic data.

Position in corporation	
Principal/CEO	3 (30%)
Manager/officer	5 (50%)
Project manager/engineer	2 (20%)
Management evaluation experience	
Yes	8 (80%)
No	2 (20%)
Area	
Plan/management	8 (80%)
Engineering	2 (20%)

Table 4. BSCI, CPI, and KPI for Korean.

BSCI	CPI	KPI
		ROIC
	Profitability	Ratio of sales cost
		Ordinary income
	Growth	Domestic sales
Finance		Overseas sales
		Debt ratio
	Stability	Rate of cash reception target accomplished
	Liquidity	Total asset turnover
	Order	Amount received from new orders
		Awards
	External customer satisfaction	Satisfaction
		Company image
		Social contribution
Customer	Internal customer satisfaction	Employee turnover rate
		Work environment & Organizational culture encouragement
	Market share	Domestic share rate
		Overseas share rate
	R&D investment	R&D cost rate
		Effectiveness against new technology development cost
	Technology competency	Applicability of self-developed technology
		Intellectual property right
Internal business process		Selling & admin. cost rate
	Operational efficiency	Compliance to guideline
		Safety rate
		Waste reuse/recycling
		Valuable resource rate
	HR development	Training cost
		Trainee satisfaction
Learning & growth	Organizational competency	Knowledge sharing level
		Employee productivity
	Information	Information competency index

a hierarchical structure. To assign weight for each indicator, interviews were conducted to experts in Korean construction firms using pair-wise comparisons method between indicators. In other words, first, the weight between BSCs indicators was analyzed on the basis of the responses of the interviews. Second, the indicators of CPIs in BSCs were analyzed. Finally, all indicators of KPIs were analyzed by using pair-wise comparisons method.

According to [18], the determinant is calculated as fuzzy vector ($\tilde{W}_i$) using Column Vector Geometric Mean Method. The fuzzy vector calculation result shows a minimum (Min $\tilde{W}_i$), arithmetic mean ($\prod_{k=1}^{n}\tilde{W}_i$), maximum value (Max $\tilde{W}_i$) and the final weight vector (W_i) estimated using the geometric mean method. In this process, calculation was performed in such a way that the sum of the weights estimated in each phase equaled one. The weights for the BSCIs, CPIs, KPIs calculated by the FD-AHP method are shown in **Table 5**. In the BSCIs, customer weight was the highest, at 0.34. For the customer CPI, the external customer satisfaction was the highest, at 0.38. For the financial aspects of the BSCI, technological competency in the internal business process, HR development and organizational competency in learning and growth showed the highest weights.

Figure 2(a) shows the BSCI weight analysis. Customers were the most important indicator, followed by Finance. This shows that the paradigm of Korean construction firms is shifting from a financial focus to a customer focus. **Figure 2(b)** represents the analyzed KPI weights. The amount received from new orders was the most important KPI, followed by the overseas market share rate, the information competency index and the domestic market share rate, which indicates that stakeholders consider order-winning competency and performance to be essential management factors. Order indicator is one of the factors which have not been considered in previous studies. In this study, it was considered to be a major

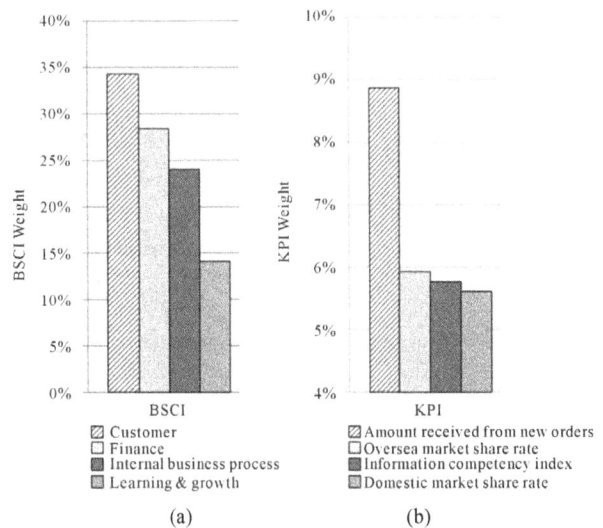

(a) (b)

Figure 2. BSCI (a) and KPI (b) weight analysis.

Table 5. Weight for BSC, CPI, and KPI.

BSCI	Weight	CPI	Weight	KPI	Weight
				ROIC	0.28
		Profitability	0.24	Ratio of sales cost	0.37
				Ordinary income	0.35
		Growth	0.18	Domestic sales	0.49
				Overseas sales	0.51
Finance	0.28	Stability	0.13	Debt ratio	0.48
				Rate of cash reception target accomplished	0.52
		Liquidity	0.13	Total asset turnover	1
		Order	0.32	Amount received from new orders	1
				Awards	0.15
		External customer satisfaction	0.38	Satisfaction	0.28
				Company image	0.38
				Social contribution	0.19
Customer	0.34	Internal customer satisfaction	0.28	Employee turnover rate	0.42
				Work environment & Organizational culture encouragement	0.58
		Market share	0.34	Domestic share rate	0.49
				Overseas share rate	0.51
		R&D investment	0.33	R&D cost rate	0.49
				Effectiveness against new technology development cost	0.51
		Technology competency	0.38	Applicability of self-developed technology	0.57
Internal business process	0.14			Intellectual property right	0.43
				Selling & admin. cost rate	0.25
		Operational efficiency	0.29	Compliance to guideline	0.21
				Safety rate	0.32
				Waste reuse/recycling	0.22
				Valuable resource rate	0.3
		HR development	0.38	Training cost	0.33
				Trainee satisfaction	0.37
Learning & growth	0.24	Organizational competency	0.38	Knowledge sharing level	0.4
				Employee productivity	0.6
		Information	0.24	Information competency index	1

indicator reflecting the characteristics of Korean construction firms. As a secondary indicator, the overseas market share rate represents the recent trend in the Korean construction industry to break into the global market.

Although the customer is the most important factor among the BSCIs, the amount received from new orders of the financial in BSCIs represents the most important indicator among the all KPIs. This can be explained from the characteristics between KPIs. Since the amount received from new orders of the financial is possible to quantify directly, the indicator can evaluate precisely from robust data. However, in the case of customer, although the external customer satisfaction in CPIs is the most important factor, since the KPIs in the factor such as awards, satisfaction, company image and social contribution are difficult to quantify the performance; the weights of the KPI level are distributed to the four indicators.

Most research has focused on identification of appropriate indicators for the management environment of the construction industry rather than estimating the importance of suitable indicators for the characteristics of the management of construction firms [10,19,20]. Major research focusing on the BSC was classified into five aspects (*i.e.*, Operational (OP), Benefits (BE), Technology/System (TS), Strategic Competitiveness (SC), and User Orientation (UO) [21-24]). Subsequently, the weight of each indicator was estimated using the AHP method. As a result, the importance was estimated in the order of OP, BE, SC, and UO, which differs from the results of Korean construction firms that emphasize the customer.

4.3. KPI Score Estimation

KPIs can be classified into quantitative indicators, such as the ROIC or debt ratio, as well as qualitative indicators, such as company image or social contribution. Quantitative indicators represent performance in numbers, while qualitative indicators need to be quantified to measure performance. For example, "cost ratio", "sales growth", and "market share" can be quantitatively presented, and order amounts, ordinary incomes and training costs can be expressed in monetary terms, while qualitative KPIs such as awards, customer satisfaction and company image cannot be directly numerically provided. Therefore, qualitative indicators need to be quantified for numerical rendering.

It is relatively easy to measure performance using financial aspects. Most of KPIs are associated with their own calculation methods. Domestic sales, overseas sales and debt ratio averages for the previous three years were compared. The rate of cash reception target accomplished was calculated by dividing the cash reception target with the actual cash reception. The target was always 100%. The amount received from new orders is the sum of orders won in a given year and is measured in monetary terms.

The awards, from the customer's point of view, are the sum of the awards accredited by governmental organizations, mass media, private organizations or overseas entities. Performance may vary, depending on specific awards, but, as the criteria is ambiguous, weights were not assigned. Customer satisfaction is measured by surveying clients and buyers, and company image is measured in terms of the ranks previously determined by professional research firms. As a part of Corporate Social Responsi-

bility (CSR), the social contribution is measured to determine the external customer satisfaction factor among CPIs. The goal of CSR is to embrace responsibility for the company's actions and encourage a positive impact through its activities on the environment, consumers, employees, communities, stakeholders and all other members of the public sphere who may also be considered as stakeholders [25]. Unlike the CSR, on the basis of donations, scholarships and activities for public benefits, the social contribution does not include production, employment, tax payments or contributions to national development. Lastly, overseas market share is limited to the aggregate overseas markets in which Korean firms are in operation, excluding comparisons with other foreign firms operating under different management conditions.

The R&D cost rate in the internal business process aspect is measured by dividing the current R&D cost with the current sales as shown in **Table 6**. Effectiveness relative to new technology development cost is the aggregation of gains rendered in monetary terms and is estimated from disciplines improved by new technologies developed on each business site, such as quality, cost, schedule, safety or environment. Intellectual property rights measured performance in terms of the aggregation of patents, new technologies and new construction methods held by the applicable corporation in a given year. In addition, compliance to guidelines is measured from surveys, and safety rate is measured using the converted safety rate calculation method.

Valuable resource rate of the learning and growth KPIs is measured in terms of the number of doctoral degrees and professional engineering certificates against the total number of employees. Training cost is measured by dividing the current training cost with the total number of employees as of the end of the current period. Trainee satisfaction and knowledge sharing level are measured using a survey, and the information competency index is determined by utilizing the information index, which is an information competency level measurement model for construction firms.

According to the weights of the scores, the final management performance varies significantly, subject to the range or estimation method of the performance measurement between the KPIs. Therefore, in order to measure balanced management performance in the MAPEC model, it is necessary to adjust the individual KPI evaluation and result calculation. As for the evaluation of the individual KPIs, a scoring interval and evaluation method which assigns 100 evaluation points, with 0 being the lowest for each item is established, and the range of performance measurement per KPI is adjusted and a scoring interval is set up. The adjusted scoring interval prevents the weights from becoming meaningless.

Table 6. Performance evaluation for finance.

BSCI	KPI	Evaluation Measurement Methods
Finance	ROIC	▷ Current net income/total asset
	Ratio of sales cost	▷ Cost of goods sold/sales × 100
	Ordinary income	▷ Operating income with addition/deduction of non-operating income/loss
	Domestic sales	▷ (Current domestic sales ÷ Average sales of recent 3 years) × 100
	Overseas sales	▷ (Current overseas sales ÷ Average sales of recent 3 years) × 100
	Debt ratio	▷ (Current debt ratio ÷ Average debt ratio of recent 3 years) × 100
	Rate of cash reception	▷ Actual/planned
	Total asset turnover	▷ Current turnover of total capital (equity + liabilities)
Customer	Amount received from new orders	▷ Sum of new orders in current period
	Awards	▷ Aggregation of awards from government organization, mass media, private organization, overseas entity (awards in the current period)
	Satisfaction	▷ Customer satisfaction rated by client and consumer on a scale of 1 - 10
	Company image	▷ Brand image studied by research firm
	Social contribution	▷ Amount of social contribution
	Employee turnover rate	▷ (Number of retirees in current period ÷ Total number of employees at the end of the current period) × 100
	Work & organization environment	▷ Employee survey
Internal business process	Domestic share rate	▷ (Orders received by company ÷ Total orders received in the market) × 100
	Overseas share rate	▷ (Overseas orders received by company ÷ Total orders received by Korean firms in the overseas market) × 100
	R&D cost rate	▷ (Current R&D cost ÷ Current sales) ×100
	Effectiveness of new technology	▷ Amount of improvement of quality, cost, schedule, safety
	self-developed technology	▷ Certification by public organizations in current period
	Intellectual property right	▷ Aggregation of patents, new technologies, new construction methods
	Selling & admin. cost rate	▷ (Current selling & admin. cost ÷ Current sales) × 100
	Compliance to guideline	▷ Average (5 point scale survey)
	Safety rate	▷ Safety rate = [(Number of death × 10 + Number of injury)/Number of permanent employees] × 100
	Waste reuse/recycling	▷ Waste for recycling ÷ Total volume of wastes
Learning & growth	Valuable resource rate	▷ (Sum of doctoral degrees and professional engineer certificates in current period ÷ Total number of employees at the end of current period) × 100
	Training cost	▷ Current training cost ÷ Total number of employees at the end of current period
	Trainee satisfaction	▷ Trainee survey
	Knowledge sharing level	▷ Cases of knowledge sharing, knowledge utilization reward (5 point scale survey)
	Employee productivity	▷ Current per-capita sales
	Information competency index	▷ IT index of construction firm ICT level measurement method model (IICI)

To adjust the scoring interval, the bottom limit of the performance is ranged along with the target value. For example, as the rate of the cash reception target accomplished rarely is less than 80%, corporations set their bottom limits to 80% or 85%, reflecting their business characteristics. In addition, as the denominator of the calculation formula for the rate of the cash reception target accomplished is the target value, the target value is always 100%. Therefore, the scoring interval of the rate of the cash reception target accomplished for most corporations can be between 100% - 80%. If a target value is to be designated, a value targeted by a given corporation for the applicable year must be selected. In the case of the amount received from new orders, the target value is the target amount to be received from new orders in the applicable year. The target value and bottom limit, set at the beginning of the year, are criteria for the performance evaluation at the end of the year. If the target value and bottom limit are not set, the scoring interval will be different than the interval in which an actual performance value exists, which may disrupt the balance of the performance evaluation. In the case of the rate of the cash reception target accomplished, if the range where the actual performance value falls is between 100% - 80%, a performance value equal to 90% of the target value can be evaluated as a 50% target accomplishment. However, if there is no scoring interval, it is evaluated as a 90% target accomplishment. Therefore, if the scoring interval is not set, the rate of the cash reception target accomplished is always positively evaluated at 80% or greater, which distorts corporate management performance scores and reduces point variance to such a small value that it renders weighting meaningless.

To evaluate the performance on a scale of 0 - 100 points, this study proposes Equations (1) and (2), which set the target value and bottom limit. Equation (1) is a function applicable to cases where the target value is greater than the bottom limit. Therefore, target value is set as the Max and the bottom limit set as the Min. If the target value is smaller than the bottom limit, as in the case of the debt ratio, Equation (2) is applicable.

$$f(x) = \frac{100}{Max - Min} x - Min\left(\frac{100}{Max - Min}\right)$$
$$= \frac{x - Max}{Max - Min} \times 100 \tag{1}$$

$$f(x)' = \frac{100}{Max - Min} + Min\left(\frac{100}{Max - Min}\right) + 100$$
$$= \frac{Max - x}{Max - Min} \times 100 \tag{2}$$

where,

Max is the greater of the target value and the bottom limit;

Min is the smaller of the target value and the bottom limit;

x is the performance value.

In this study, as the score is calculated by Equations (1) and (2), the scoring interval and distribution in all KPIs are distributed equally between 0 and 100. However, as the actual performance value may exceed the

target or fall below the bottom limit. It would not fit within the range of 0 - 100 points. Therefore, if the actual performance value is below 0 or greater than 100 points, it is calculated as 0 and 100 points, accordingly.

5. Case Study

5.1. Overview

To verify the applicability of the MAPEC developed in this study, actual corporate performance values were entered, and performance was evaluated. The two of the 2009 top 30 major construction firms ranked in terms of construction competency were selected. **Table 7** shows the profiles of corporation A and B that were selected for this case study. Corporation A has experienced growth for the previous four years. In addition, Corporation A's construction competency valuation increased from 1.1 trillion won to 1.6 trillion won in 2009, and it posted an operating income of 105.7 billion won in 2008 and 70.6 billion up to the third quarter of 2009, which shows that both the corporation's growth potential and profitability are high. In other words, Corporation A has been performing fairly positively in terms of the financial aspects. Corporation B is a major construction firm posting 7 trillion won in sales on average in last three years. It posted 8.2 trillion won in construction competency valuation in 2009, and its current net income decreased from 990 billion in 2007 to 80 billion in 2009.

5.2. Practical Application

In order to measure the performances of Corporations A and B against their targets, they were examined in accordance with the criteria applicable to each KPI item, and the weighted KPI scores were calculated as shown in **Table 8**. The KPI scores were weighted and converted to scores for each performance aspect. The performance aspect scores were then weighted in accordance with criteria specific to each aspect and converted to BSC scores. Comprehensive management performance was calculated by multiplying the BSCI score with the BSCI weight.

Corporation A received 100 points for the 10 KPIs of ROIC, sales growth, amount received from new orders, awards, customer satisfaction, employee turnover, intellectual property rights, safety rights, knowledge sharing level and information competency index. This corporation posted various points, from 0 to 90, for the other KPIs, with an average KPI score of 53.5 points.

Corporation B received 100 points for the 10 KPIs of social contribution, employee turnover, R&D cost rate, intellectual property rights, compliance to guideline, safety rates, valuable resource rates, training costs, knowledge sharing level and information competency

Table 7. Overview of cases.

Classification	"A" Case				"B" Case		(In: 100 M KRW)	
	2006	2007	2008	2009	2006	2007	2008	2009
Construction competency	11,553	11,903	12,360	16,082	65,600	76,635	89,272	82,572
No. of employees	1918	2197	2404	2292	3205	3420	3651	4811
Credit rating	BBB	BBB	BBB	BBB	A	A-	A-	A-
Sales	14,030	16,063	23,493	16,724	57,291	61,554	65,915	70,974
Cost of goods sold	12,938	14,795	21,308	15,304	48,336	52,618	58,866	65,365
Sales income	1092	1268	2185	1420	8955	8936	7049	5609
Selling & admin. cost	1005	1073	1128	706	2667	3117	3469	3414
Operating income	87	195	1057	714	6288	5819	3580	2195
Non-operating income	728	957	369	224	3671	9937	6288	4754
Non-operating cost	559	1552	1047	619	3652	1986	5842	5620
Income tax	99	−74	141	85	1924	3853	1514	529
Current net income	157	−326	238	234	4383	9917	2512	800
Equity capital	126	126	126	126	16,965	16,286	16,286	16,286
Total asset	1501	1608	2015	2187	60,847	68,492	94,455	88,410
Total liabilities	10,519	11,856	14,738	15,826	33,109	37,502	61,246	57,859
Total capital	4492	4228	5415	6043	27,738	30,990	33,209	30,551
Debt ratio	234.2%	280.4%	272.2%	261.9%	119.4%	121.0%	184.4%	189.4%
Dependency on borrowing	38.8%	32.7%	30.0%	30.2%	14.1%	14.0%	26.2%	30.0%
Times interest earned	0.18 times	0.43 times	2.0 times	1.64 times	8.3 times	10.9 times	2.0 times	1.2 times

index (**Table 7**). For the other KPIs, it posted points in between 0 to 90, with an average score of 67.4 points, which is 8 points higher than Corporation A.

The analysis of the two corporations found that Corporation A received 53.5 points for comprehensive management performance, and B scored 67.4 points. While Corporation A's average KPI score was 4.4% higher than the comprehensive management performance, Corporation B's comprehensive management performance was 1.5% higher than its average KPI score, which is attributable to the fact that Corporation B scored higher in the more important items, indicating that it performed better in terms of the more important management items.

In the case of Corporation A, the customer and learning & growth factors in BSCIs represent relatively lower than the Corporation B. Especially, the lower customer factor had impact negatively on overall management performance. Therefore, although the A and B cases are similar in the financial factor, the Corporation A should

be adjusted the strategies in the customer and learning & growth factors to improve management performance in the long term. In the case of the Corporation B, since financial factor is lower than the A case, the B should be concentrated on the KPIs of financial factor.

6. Discussion and Conclusions

Most corporations focus on financial indicators, such as current net income, ROI, and ROE, which are easy to measure and capable of showing management performance in quantifiable forms. However, it is difficult to evaluate the potential of a corporation by relying only on the financial aspects in management performance evaluations. Accordingly, this research proposes the MAPEC in an effort to ensure a systematic performance measurement by quantifying the evaluation results of qualitative indicators compatible with the characteristics of the construction industry environment. The results drawn from this research are as follows.

Table 8. "A", "B" case management performance.

Performance A	Performance B	BSC	BSC Score A	BSC Score B	Area	CPI Score A	CPI Score B	KPI	KPI Score A	KPI Score B
								ROIC	100.0	10.8
					Profitability	54.9	29.5	Ratio of sales cost	19.2	50.0
								Ordinary income	56.4	22.7
					Growth	13.1	54.2	Domestic sales	26.8	53.8
		Finance	57.2	52.4				Overseas sales	0.0	54.5
								Debt ratio	38.1	39.6
					Stability	24.0	50.7	Rate of cash reception target accomplished	11.0	61.0
					Liquidity	50.0	26.7	Total asset turnover	50.0	26.7
					Order	100.0	79.8	Amount received from new orders	100.0	79.8
								Awards	100.0	94.0
					External satisfaction	79.2	85.5	Satisfaction	100.0	96.7
								Company image	75.0	66.7
								Social contribution	40.0	100.0
		Customer	56.2	74.0	Internal satisfaction	53.6	65.2	Employee turnover rate	100.0	100.0
								Work environment & Organizational culture encouragement	20.0	40.0
53.5	67.4				Market share	32.7	68.3	Domestic share rate	66.7	70.0
								Overseas share rate	0.0	66.7
					R&D investment	28.0	49.0	R&D cost rate	57.1	100.0
								Effectiveness against new technology development cost	0.0	0.0
					Technology competency	71.5	88.6	Applicability of self-developed technology	50.0	80.0
		Internal business process	53.8	65.2				Intellectual property right	100.0	100.0
								Selling & admin. cost rate	35.7	0.0
					Operational efficiency	59.8	53.0	Compliance to guideline	90.0	100.0
								Safety rate	100.0	100.0
								Waste reuse/recycling	0.0	0.0
								Valuable resource rate	40.0	100.0
					HR development	35.3	81.5	Training cost	28.6	100.0
								Trainee satisfaction	37.5	50.0
		Learning and growth	45.4	77.0	Organizational competency	84.2	58.0	Knowledge sharing level	100.0	100.0
								Employee productivity	73.7	30.0
					Information	0.0	100.0	Information competency index	0.0	100.0

First, as the MAPEC proposed herein applied the FD-AHP method to weigh each evaluation item and estimated performance indicators in pair-wise comparison, accurate and systematic management performance measurement results were obtained.

Second, the analysis of the BSCI weights showed that the customer is the most important variable, followed by finance. Thus, major Korean construction firms are shifting their focus from revenue creation to customer service, which reveals a significant difference from the preceding studies, which emphasized operational levels (Document transfer and handling, Coordination and communication, Response times, Support alliance relationships, Decision making, Reporting).

Third, the analysis of the KPI weights showed that the amount received from new orders is the most important KPI, followed by overseas market share, information competency index, and domestic market share. This indicates those with hands-on responsibility for corporate management regard order-winning capabilities and performance as important management factors.

Fourth, a management performance measurement model for Korean construction firms was developed and actual management planning, execution and feedback processes were analyzed. In addition, preceding studies introduced the BSC and focused only on its concept, which resulted in errors and mistakes in its actual application. This model, however, presents a generic process to be followed, allowing Korean construction firms to apply the model to their actual operations.

This research used a survey to confirm that the management of Korean construction firms believes that customer satisfaction and brand image have more impact on their corporate management than did management performance in terms of financial statements. In addition, the MAPEC proposed herein not only assigns weights appropriate for the Korean construction industry environment, but also suggests further details relevant to management performance, including KPI selection and evaluation methods, delivering a close-to-standard management performance measurement model for construction firms.

The performance measurement model developed herein can be utilized when analyzing the cause of deficiency (vulnerability) for each BSCI-CPI-KPI item beyond a simple measurement of the management performance of each corporation. In other words, the performance measurement results can be fed back into strategies and plan to shed light on issues, reflect on management plans for subsequent years and modify mid to long-term strategies. Furthermore, the model implementation process suggested herein can be applied to the implementation of management performance measurement models not only for other big corporations, but also small

and medium-sized construction firms.

Although this research proposes KPIs appropriate for Korean construction firms, it did not consider the complex and dynamic interrelationships among different KPIs. Therefore, it is necessary to analyze and systematically substantiate the dynamic relationships that influence different KPIs.

REFERENCES

[1]   P. F. Drucker, "The Information Executives Truly Need," *Harvard Business Review*, Vol. 73, No. 1, 1995, pp. 54-62.

[2]   E. K. Zavadskas, Z. Turskis and J. Tamosaitiene, "Risk Assessment of Construction Projects," *Journal of Civil Engineering and Management*, Vol. 16, No. 1, 2010, pp. 33-46.

[3]   C. D. Ittner and D. F. Larcker, "Innovations in Performance Measurement: Trends and Research Implications," *Journal of Management Accounting Research*, Vol. 10, No. 2004, pp. 205-238.

[4]   R. F. Cox, R. R. A. Issa and D. Ahren, "Management's Perception of Key Performance Indicators for Construction," *Journal of Construction Engineering and Management*, Vol. 129, No. 2, 2003, pp. 142-151.

[5]   M. Kagioglou, R. Cooper and G. Aouad, "Performance Management in Construction: A Conceptual Framework," *Construction Management and Economics*, Vol. 19, No. 2001, pp. 93-106.

[6]   H. S. Cha and T. K. Kim, "Developing Measurement System for Key Performance Indicators on Building Construction Projects," *Korea Institute of Construction Engineering and Management*, Vol. 9, No. 4, 2008, pp. 120-131.

[7]   I. H. Yu, K. R. Kim, Y. Jung and S. Chin, "Analysis of Quantified Characteristics of the Performance Indicators for Construction Companies," *Korea Institute of Construction Engineering and Management*, Vol. 7, No. 4, 2006, pp. 154-164.

[8]   K. H. Kim, I. H. Yu, D. W. Shin and K.R. Kim, "Evaluation of Performance Measurement System Alternatives for the Construction Companies," *Architectural Institute of Korea*, Vol. 21, No. 6, 2005, pp. 97-105.

[9]   R. S. Kaplan and D. P. Norton, "The Balanced Scorecard—Measures That Drive Performance," *Harvard Business Review*, Vol. 70, No. 1, 1992, pp. 71-79.

[10] W. J. Jung, I. H. Yu, K. R. Kim and D. W. Shin, "Analysis of the Weights of Performance Measurement Index According to the Size of Construction Companies," *Architectural Institute of Korea*, Vol. 21, No. 8, 2005, pp. 121-129.

[11] A. J. Shrnhur, O. Levy and D. Dvir, "Mapping the Dimensions of Project Success," *Project Management Journal*, Vol. 28, No. 2, 1997, pp. 5-13.

[12] C. S. Lim and M. Z. Mohamed, "Criteria of Project Success and Exploratory Re-Examination," *International Jour-*

nal of Project Management, Vol. 17, No. 4, 1999, pp. 243-248.

[13] R. Atkinson, "Project Management: Cost, Time and Quality, Two Best Guesses and a Phenomenon, Its Time to Accept Other Success Criteria," *International Journal of Project Management*, Vol. 17, No. 6, 1999, pp. 337-342.

[14] M. Kagioglou, R. Cooper and G. Aouad, "Performance Management in Construction: A Conceptual Framework," *Construction Management and Economics*, Vol. 19, No. 1, 2001, pp. 85-95.

[15] D. H. Lee, S. K. Kim and D. H. Shin, "A Correlation Analysis between the Change of Managerial Environment and the Business Performance of Domestic Construction Firms." *Journal of the Korea Institute of Building Construction*, Vol. 9, No. 1, 2009, pp. 111-121.

[16] S. O. Ogunlana1, H. Li and F. A. Sukhera, "System Dynamics Approach to Exploring Performance Enhancement in a Construction Organization," *Journal of Construction Engineering and Management*, Vol. 129, No. 5, 2003, pp. 528-536.

[17] A. Brown and J. Adams, "Measuring the Effect of Project Management on Construction Outpus: A New Approach," *International Journal of Project Management*, Vol. 18, No. 5, 2000, pp. 327-335.

[18] Y. C. Liu, "Application of the Fuzzy Delphi Analytic Hierarchy Process on Rock Mass Classification," National Cheng Kung University, Tainan City, 2002.

[19] H. A. Bassioni, A. D. F. Price and T. M. Hassan, "Building a Conceptual Framework for Measuring Business Performance in Construction: An Empirical Evaluation," *Construction Management and Economics*, Vol. 23, No. 5, 2005, pp. 495-507.

[20] S. H. Chen and H. T. Lee, "Performance Evaluation Model for Project Managers Using Managerial Practices," *International Journal of Project Management*, Vol. 25, No. 6, 2007, pp. 543-551.

[21] C. Chiu, "A Case-Based Customer Classification Approach for Direct Marketing," *Expert Systems with Application*, Vol. 22, No. 2, 2002, pp. 163-168.

[22] V. T. Luu, S. Y. Kim and T. A. Huynh, "Improving Project Management Performance of Large Contractors Using Benchmarking Approach," *International Journal of Project Management*, Vol. 26, No. 7, 2008, pp. 758-769.

[23] R. A. Stewart and S. Mohamed, "Utilizing the Balanced Scorecard for IT/IS Performance Evaluation in Construction," *Construction Innovation*, Vol. 1, No. 3, 2001, pp. 147-163.

[24] N. K. Acharya, Y. D. Lee and J. K. Kim, "Critical Construction Conflicting Factors Identification Using Analytical Hierachy Process," *KSCE Journal of Civil Engineering*, Vol. 10, No. 3, 2006, pp. 165-174.

[25] D. Wood, "Corporate Social Performance Revisited," *The Academy of Management Review*, Vol. 16, No. 4, 1991, pp. 34-42.

6

Modeling Energy Generation by Grid Connected Photovoltaic Systems in the United States

Robert E. Steffen, Sung Joon Suk[*], Yong Han Ahn, George Ford

Department of Construction Management, Western Carolina University, Cullowhee, USA.

ABSTRACT

This article presents the results of an analysis of hourly data obtained from forty-three photovoltaic (PV) systems installed in North America. Energy data collected from these systems were organized according to monthly output in an effort to identify factors which are effective in predicting energy generation. Independent variables such as system capacity, shading, longitude, latitude, seasonal variation, and orientation were considered. Multiple regression analysis was used to quantify the kilowatt-hours that can be expected from a change in the independent variables. Results show that all six independent variables are significant predictors which can be used in a regression model to estimate system output with a high level of confidence. The analysis shows that approximately 83% of the variation in the amount of energy generated monthly by the forty-three solar panels is explained by the independent variables and the derived equation. Results of the study may prove helpful to solar panel system users who may need to consider less than optimum conditions during a PV panel installation and service life.

Keywords: Solar Energy; Photovoltaic System; Regression Analysis

1. Introduction

The performance and efficiency of PV systems have increased dramatically over the last decade [1-3] while installation and maintenance costs of systems have declined. From 2008 to 2012 solar panel prices have decreased by 25% in the United States from approximately $4/Watt, with many currently installed systems below $3/Watt [4]. Cost savings are a result of many factors including improved materials and manufacturing, economies of scale, improved financing and buy-back options, as well as federal and state government tax credit programs [4]. Other factors are weighed when procuring a PV system, such as the known solar insolation of a given area [5], cloud cover, energy payback associated with regional energy provider(s), individual building factors such as energy demands [6], and environmental effects including air pollution and dust [7,8]. Compared to conventional energy options and considering multiple geographic and buy-back regions, PV generated electricity remains expensive, although prices are falling due to government incentives and the rapid expansion of the industry [9,10].

Energy generation information should be available for potential PV system owners who wish to compare PV output to conventional energy supplier output. There currently exists a need for potential PV investors to collect and compare current energy generation data with a known degree of accuracy. Moreover, the decision to invest in a PV system often depends upon monthly-generated income as well as the payback period. Thus data should be available between and within prospective PV system installation regions. Energy generation data obtained from recently installed systems could greatly influence a residential or commercial owner's decision to invest in a PV system. However, currently available software used to predict PV energy generation, such as PV Watts [11], has limitations. Known limitations include climate factors, pollutants, unknown maintenance costs, shading, and/or simulation errors. Therefore it is of great importance to find the best method to accurately collect data and subsequently predict electricity generation from known PV systems. Collected information can then be incorporated within the decision making process related to investment and installation. In this regard the purpose of the current study is to develop a predictive model for estimating PV output based on available energy generation data within a known confidence level.

2. Literature Review

Although studies have attempted to identify predictive models for electricity generation capacity from PV systems, uncertainties and limitations point to a need for additional research. Thevenard and Pelland [12] conducted a study that estimated the uncertainty in long-term PV yield prediction through statistical modeling of a hypothetical 10 megawatt (MW) PV system in Toronto, Canada. The study found uncertainties including: 3.9% uncertainty for year-to-year climate variability; 5% for long-term average horizontal insolation; 3% for power rating of modules; 2% for losses due to dirt and soiling; 1.5% for losses due to snow; and 5% for other sources of error [12]. Oozeki [13] studied loss factors of PV system operations and classified six kinds of system losses including shading effect, losses due to incident angle, loading mismatch, efficiency decrease due to temperature effects, inverter losses, as well as other losses. Oozeki [13] also developed a calculation method to predict the energy production of a PV system using irradiance-domain integrals and the definition of a statistical moment.

Mau and Jahn [14] analyzed long-term performance and reliability issues of 21 selected PV systems in five different countries in Europe and in Japan. Similarly, Marion et al. [15] conducted a study to identify performance parameters for grid-connected PV systems. These performance parameters were discussed for their suitability in providing desired information for PV system design and performance evaluation, and were demonstrated for a variety of technologies, designs, and geographic locations. Hong et al. [16] estimated the loss ratio of solar PV electricity generation through stochastic analysis.

Additional studies of energy generation data obtained from installed PV systems are needed in order to guide prospective owners through the entire PV installation and energy generation process.

3. Research Methodology

The current research analyzed 505 energy output data points generated by numerous panels installed in the United States. The data were collected from the PV Ouput website (http://pvoutput.org), a publicly accessible data bank containing individual members PV energy generation data. Upon reviewing the output of 285 solar panel systems in the United States, a total of 43 systems were selected for analysis on the basis of data set completeness for the years 2011 and 2012. **Table 1** provides descriptive statistics including energy generated monthly, system capacity (size), orientation, and shading.

Overall, this study used data from the solar panels installed across the United States, although more solar systems were found on the west coast and in the northeast. **Figure 1** shows the installation locations of the 43 solar panel systems.

Multiple regression analysis was used to identify relationships between the amount of energy generated monthly by the studied solar PV systems and various potential predictors: system capacity, installed orientation, location, shading status, and season. The characteristics and definitions of the variables included in the regression analysis are shown in **Table 2**. To alleviate multi-collinearity, three continuous independent variables were transformed by mean centering. The reference groups for each categorical variable are south for orientation, no for shading status, and summer (or not summer) for season.

4. Results

Standard multiple regression was conducted with the variables described in **Table 2**. Relevant statistical results of these variables are summarized in **Tables 3-5**. As reported in **Table 3**, an analysis of variance for the regression model indicates that the Multiple R for the regression model was statistically significant ($F(8,496) = 307.72$, $p < 0.001$). As shown in Table 4, the regression model accounts for 83.2% of the variation in the observed monthly energy generation values in the sampled

Table 1. Descriptive statistics of forty-three installed solar panels.

	Min.	Max.	Average
Monthly Energy Generation (kWh)	89.7	1608.0	684.1
System Capacity (kW)	1.38	11.04	5.65
Orientation	South (27)		Other (16)
Shading	No (20)		Yes (23)

Figure 1. Geographic locations of forty-three individual solar panel systems (Map Data ©2103 Google INEGI Maplink TeleAtlas).

Table 2. Characteristics and definitions of variables.

		Characteristic	Definition & Dummy Coding
Amount of energy generated monthly by solar panels	Dependent Variable	Continuous	
System capacity		Continuous	Mean (5.65) centered
Orientation		Categorical	Orientation D1: South (0) or Other orientations (1)
Location — Latitude	Independent Variable	Continuous	Mean (38°50'N) centered
Location — Longitude		Continuous	Mean (95°40'W) centered
Shading status		Categorical	Shading D1: No (0) or Yes (1)
Season		Categorical	Season D1 Season D2 Season D3	• Spring (1,0,0): Mar., Apr., May • Summer (0,0,0): Jun., Jul., Aug. • Fall (0,1,0): Sept., Oct., Nov. • Winter (0,0,1): Dec., Jan., Feb.

Table 3. ANOVA results.

	Sum of Squares	Degree of Freedom	Mean Square	F	Sig.
Regression	51246522.0	8	6405815.3	307.7	0.000
Residual	10325095.5	496	20816.7		
Total	61571617.5	504			

Table 4. Model summary.

R	R^2	Adjusted R^2
0.912	0.832	0.830

systems ($R^2 = 0.832$), and an estimated 83% of the variance in output projected for the population of similar systems (R^2 *adj.* = 0.83).

Based on the results shown in **Table 5**, a regression equation was found, where the predicted monthly energy generation can be found from the equation: Predicted Energy generation = 912.584 + 124.743 (System Capacity – 5.63) – 29.127 (Orientation D1) – 9.027 (Latitude – 38°50'N) + 2.743 (Longitude – 95°40'W) – 50.845 (Shading D1) – 67.054 (Season D1) – 261.658(Season D2) – 394.221 (Season D3). As shown by the t values and the corresponding significance values, all of the slopes are statistically significant (*i.e.*, we can conclude at the 0.05 level that all slopes are not zero). Hence, all the independent variables contribute in predicting the amount of energy generated monthly by the solar panels included in the sample. Since none of the independent variables has a variance inflation factor (VIF) greater than five, there are no apparent multi-collinearity problems [17]; in other words, there is no variable in the model that measures the same relationship/quantity as is measured by another variable. Moreover, the fitted model was found to not violate other basic assumptions required in a valid regression model. Because the coefficients of the inde-

pendent variables are all significant at the 0.05 level of significance, the average amount of energy generated monthly by the given PV system set may be adjusted as follows by each independent variable, holding other variables constant:

• Monthly energy generation will increase by approximately 125.3 (kWh) for each 1 (kW) increase in the system capacity of the solar panels.
• For the given geographic region, solar panels facing south will generate as much as 29.1 (kWh) more energy than solar panels facing other orientations.
• Regarding locations within the United States, power generation will be larger in solar panels installed in the western or southern United States than in the eastern or northern United States. This energy generation will increase by 9.0 (kWh) for each 1 degree decrease in latitude and will increase by 2.7 (kWh) for each 1 degree increase from center longitude.
• Shading will reduce the amount of energy generation as much as 50.85 (kWh).
• As expected, solar panels will generate more energy in summer than in other seasons; 67.05, 261.7, and 394.2 more energy (kWh) in summer than in spring, fall, and winter, respectively.
• Regarding the interpretation of the numeric constants in the equation, a solar panel will generate approximately 912.6 (kWh) energy when the system capacity is 5.63 (kW), the panels are facing south, are located at the latitude of 38°50'N and the longitude of 95°40'W, are not shaded, and during the summer season.

Table 5. Regression coefficients of independent variables.

	Slope	Standard Error	Standardized Coefficients	t	Sig.	Collinearity Statistics	
						Tolerance	VIF
Constant	912.584	15.260		59.802	0.000		
System Capacity	125.264	2.930	0.795	42.756	0.000	0.978	1.023
Orientation D1	−29.127	14.532	−0.039	−2.004	0.046	0.899	1.113
Latitude	−9.027	1.546	−0.113	−5.839	0.000	0.901	1.110
Longitude	2.743	0.340	0.154	8.070	0.000	0.924	1.082
Shading D1	−50.845	13.242	−0.073	−3.840	0.000	0.941	1.063
Season D1	−67.054	18.152	−0.081	−3.694	0.000	0.695	1.440
Season D2	−261.658	17.244	−0.340	−15.174	0.000	0.675	1.483
Season D3	−394.221	18.811	−0.458	−20.957	0.000	0.707	1.414

While the slope coefficients all contribute to the prediction of PV output, the relative importance can be gauged by the Beta, or Standardized Coefficient values given in **Table 5**. Betas with larger absolute values are more important than those with smaller absolute values given that they are in standard deviation units instead of the varying units of the original independent variables.

5. Discussion and Conclusion

This study examined the relationships between the amount of energy generated monthly by a known set of solar panels and various predictors. Based on a total of 505 energy output data points generated monthly by forty-three solar panels installed in the United States, a regression equation was derived for predicting photo voltaic energy generation of PV systems with a system capacity between 1.4 and 11.0 kW. The monthly energy generation, $\hat{y}$, can be predicted from the equation:

$$\hat{y} = 912.584 + 124.743X_1 - 29.127X_2 - 9.027X_3$$
$$+ 2.743X_4 - 50.845X_5 - 67.054X_6$$
$$- 261.658X_7 - 394.221X_8$$

where:
- $\hat{y}$ = Predicted monthly photo voltaic energy generation;
- X_1 = System Capacity − 5.63 (in Kw);
- X_2 = Orientation (0 for South, 1 for other orientations);
- X_3 = Latitude − 38°50'N;
- X_4 = Longitude − 95°40'W;
- X_5 = Shading (No (0) or Yes (1));
- X_6 = Season D1 (Spring, 0 or 1);
- X_7 = Season D2 (Fall, 0 or 1);
- X_8 = Season D3 (Winter, 0 or 1).

By entering a value for each of the variables X1-X8, users can obtain energy predictions for prospective installations within the range of the systems sampled in this study. In such applications, the regression model developed herein explains over 83% of the variation typically seen in solar PV systems in the US. Moreover, given that the overall model is statistically significant, it is reasonable to assume that the given regression equation can accurately predict the PV output with the parameters described above.

The equation is presented as a practical, simple prediction of small-scale PV system energy generation and an explanation of critical considerations required for solar PV installations. This model will prove helpful to individuals with little experience in solar power systems who may be considering less than optimum conditions for a PV panel installation.

Proper system capacity may be determined depending on a required amount of energy generated monthly and yearly, as well as a given system location, orientation, and shading status. To generate equivalent energy amounts, PV systems in the eastern or northern United States should have larger system capacities than western or southern areas. Additionally, solar panels not facing directly south, or shaded, should have a larger system capacity in order to produce an equivalent output. Additionally, the Beta values produced in the regression analysis provide guidance regarding the importance of the criteria that can be used to further refine the decision making process. Of primary concern is the system capacity. Next, seasonal variation substantially impacts PV output, followed by longitude and then latitude. It is interesting to note that shading and orientation do not appear as variables of central importance. In fact, orientation has a significance value just slightly lower than the 0.05 probability threshold. These relatively low values for orientation and shading may be attributable to relatively small amounts of variation as well as general installation considerations for PV systems. For these variables, while in general there may be small departures from ideal installations, both the installer and the owner have vested interests in approximating ideal conditions. It is difficult to imagine either party tolerating large amounts of shading, or large departures from southerly orientation, so as to maximize solar exposure. Therefore, given the expense and effort required for the installation of systems like those sampled in this study, one would expect that care has been taken to obtain good solar exposure.

The regression model developed in this study can be beneficially utilized by perspective owners of PV solar systems to evaluate the claims of manufacturers and installers relative to a model prediction based from a sample of previously installed systems. Such a comparison can assist perspective owners to better assess both returns on their investment (ROI) and payback periods associated with proposed systems.

6. Limitations and Future Work

This study did not address the variation in PV output associated with system manufacturers, actual inverter efficiencies, and system downtime. Systems produced by different manufacturers vary with respect to quality and efficiency. Likewise, the quality of inverter efficiencies varies widely. Further, the datasets used did not consider discontinuous timelines which would account for system downtime, often attributed to panel damage or failure, inverter failure, weather damage, switch failures, or operator error. Additionally, isolated regions of non-typical cloud cover and solar insolation were not accounted for. Similarly, certain longitudes in the US contained no data sets with continuous energy generation output.

The model generally considered only smaller (1.4 to 11 Kw), grid connected PV systems in capacity sizes common to residential, privately owned PV systems. In comparison with isolated systems, grid-connected systems would be expected to have higher efficiencies due to load balancing. All the energy generated by the PV systems in this study is assumed to be used in grid connected systems where there is always a load greater than the power produced. Isolated, residential systems operating off the grid may utilize all available generated capacity, but must be load balanced.

REFERENCES

[1] J. L. Lee and S. Y. Kim, "Enhanced Efficiency of Organic Photovoltaic Cells with Sr_2SiO_4:Eu2+ and $SrGa_2S_4$: Eu2+ Phosphors," *Organic Electronics*, Vol. 14, No. 4, 2013, pp. 1021-1026.

[2] X. Zhang, X. Zhao, S. Smith, J. Xu and X. Yu, "Review of R&D Progress and Practical Application of the Solar Photovoltaic/Thermal (PV/T) Technologies," *Renewable and Sustainable Energy Reviews*, Vol. 16, No. 1, 2012, pp. 599-617.

[3] P. Biwole, P. Eclache and F. Kuznik, "Phase-Change Materials to Improve Solar Panel's Performance," *Energy and Buildings*, Vol. 62, 2013, pp. 59-67.

[4] A. M. Paudel and H. Sarper, "Economic Analysis of a Grid-Connected Commercial Photovoltaic System at Colorado State University-Pueblo," *Energy*, Vol. 52, 2013, pp. 289-296.

[5] P. Denholm and R. Margolis, "The Regional Per-Capita Solar Electric Footprint for the United States," National Renewable Energy Laboratory, Golden, 2007.

[6] S. Danov, J. Carbonell, J. Cipriano and J. Martí-Herrero, "Approaches to Evaluate Building Energy Performance from Daily Consumption Data Considering Dynamic and Solar Gain Effects," *Energy and Buildings*, Vol. 57, 2013, pp. 110-118.

[7] J. K. Kaldellis and A. Kokala, "Quantifying the Decrease of the Photovoltaic Panels' Energy Yield Due to Phenomena of Natural Air Pollution Disposal," *Energy*, Vol. 35, No. 12, 2010, pp. 4862-4869.

[8] K. A. Moharram, M. S. Abd-Elhady, H. A. Kandil and H. El-Sherif, "Influence of Cleaning Using Water and Surfactants on the Performance of Photovoltaic Panels," *Energy Conversion and Management*, Vol. 68, 2013, pp. 266-272.

[9] G. Boyle, "Renewable Energy: Power for A Sustainable Future," Oxford University Press, Oxford, 2004.

[10] F. Cucchiella and I. D'Adamo, "Estimation of the Energetic and Environmental Impacts of a Roof-Mounted Building-Integrated Photovoltaic Systems," *Renewable and Sustainable Energy Reviews*, Vol. 16, No. 7, 2012, pp. 5245-5259.

[11] NREL, "PVWatts Software," National Renewable Energy Laboratory, Golden, 2013. http://rredc.nrel.gov/solar/calculators/pvwatts/version1

[12] D. Thevenard and S. Pelland, "Estimating the Uncertainty in Long-Term Photovoltaic Yield Predictions," *Solar Energy*, Vol. 91, 2013, pp. 432-445.

[13] T. Oozeki, T. Izawa, K. Otani and K. Kurokawa, "An Evaluation Method of PV Systems," *Solar Energy Materials and Solar Cells*, Vol. 75, No. 3-4, 2003, pp. 687-695.

[14] S. Mau and U. Jahn, "Performance Analysis of Grid-Connected PV Systems," 21*st European Photovoltaic Solar Energy Conference*, Dresden, 4-8 September 2006.

[15] B. Marion, J. Adelstein, K. Boyle, H. Hayden, B. Hammond, T. Fletcher, B. Canada, D. Narang, D. Shugar, H. Wenger, A. Kimber, L. Mitchell, G. Rich and T. Townsend, "Performance Parameters for Grid-Connected PV Systems," 31*st IEEE Photovoltaics Specialists Conference and Exhibition*, Lake Buena Vista, 3-7 January 2005,

pp. 1601-1606.

[16] T. Hong, C. Koo and M. Lee, "Estimating the Loss Ratio of Solar Photovoltaic Electricity Generation through Stochastic Analysis," *ICCEPM*, Garden Grove, 9-11 January 2013, pp. 389-399.

[17] R. O'Brien, "A Caution Regarding Rules of Thumb for Variance Inflation Factors," *Quality & Quantity*, Vol. 41, No. 5, 2007, pp. 673-690.

7

Framework of Construction Procedure Manuals for PMIS Implementation

Boong Yeol Ryoo

Department of Construction Science, Texas A&M University, College Station, USA.

ABSTRACT

Primarily project procedure manuals are intended to enforce company polices or procedures. These manuals are important pieces for successful project management in the construction industry, because construction projects are operated virtually. Even though subcontractors are responsible for field works, a general contractor is still responsible for providing quality project management services for project owners. The more subcontractors involved, the harder it is to monitor and control them, due to the different management processes and procedures they use. More than 36 procedure manuals in the building and industrial construction were reviewed to create process maps of various management services. According to surveys, larger contractors have a broader use of procedure manuals than smaller contractors. The full-scale manuals cover project administration, schedule, cost, contract administration, quality, or safety to home office operations. The small-scale manuals cover from site mobilization, to startup and closeout. Small-scale operation manuals are often used by medium to small contractors. This paper presents suggestions for integrated procedure manuals for construction management firms based on studies of multiple construction procedure manuals. A framework for integrating procedure manuals is presented. The proposed framework can be used to keep uniformity across management procedures and phases. In addition, it can be used to implement a project management system. It can be used to forecast or evaluate management activities through replicability of the management responses.

Keywords: Procedure Manuals; Action Words; Project Management System

1. Introduction

1.1. Background

As part of their response to a request for proposal (RFP), a general contractor submits a set of project management manuals to the owner as an element of the contract documents. These are used to determine the level of management competitiveness of the general contractor. The manuals define necessary management tasks, processes, and procedures with included forms for those procedures. It is an essential piece for successful project management, because, while the management team and the subcontractors are connected through the project, each of them has different management control.

In construction, only a small number of general contractors actively implement project management procedures to describe how they manage building construction projects. The majority tend to only use the documents and forms in the manual to track project issues, manage project budgets, and document issues for legal reasons.

For those who do actively use these procedures they are given an advantage. Because most of them are in a specialized limited number of construction sectors, it helps them to develop a management plan, examine past performance or improve management performance and methods if they have a written procedure manual [1].

The specialization of construction companies and a lack of standardization in construction project management hinder effective communication and documentation between contractors. The main issue in a relationship between contractors is neither of the parties wants to give up their management procedures, because both parties have built their entire systems on these procedures and there is difficulty in changing over to another system.

The end result of this lack of standardization and minority of companies that actively manage with procedures is that there are insufficient procedures for management tasks and commitments. This often leaves managers no choice but to rely on their past experience and individual knowledge to deal with managerial challenges.

In the short run this usually proves to be sufficient but this non-replicable style makes it difficult to study project performance to determine effective procedures. Also, a non-replicable style makes it difficult to have predictability for results, or incidents in a project.

While there are several professional groups in the construction industry there is still a lack of standards that could be applied in this arena. For example the Construction Management Association of America (CMAA) has released a list of standard services of construction management (CM) in seven management areas [2]. This is representative of the nature of the industry, constructors know what needs to be delivered; they just have a difference of opinion on how to deliver the final product. Because of this, neither specific processes nor procedures are suggested. This lack of specificity leaves it to the individual general contractors to be responsible for creating instructions for construction managers. This responsibility sometimes falls to the individual construction manager, who is given a set of general rules and deliverables and is left to come up with their own method of delivering within those rules.

To aid in construction management, project management information systems (PMIS) have been adopted as a project controlling and monitoring tool. PMISs enable project managers to collect, distribute, measure, and process project data. Initially, large contractors had a tendency to use a custom designed PMIS for their management systems. Recently, however, they have started to adopt an off-the-shelf PMIS and adjusting their management systems to the PMIS. Mainly this has been done due to the expense associated with a customized software system as opposed to one that is mass produced.

Because the majority of the companies use non replicable styles, procedures have become unique to not only each company but to each manager in that company. Thus, synchronizing an off-the-shelf PMIS and procedure manuals are a challenge because this method of operation [3].

1.2. Purpose of the Study

The purpose of this paper is to propose a framework for integrated project management in the building industry. By studying processes and methods of existing procedure manuals, a base structure for project control and monitoring procedures is proposed.

Guaranteed Maximum Price CM (GMPCM) projects are an intended target because their management responsibility includes both project management and construction operations. This gives them an opportunity to manage, perform and report on the projects they are involved in. Another feature of this framework is a list of management tasks mapped to illustrate various types of ma-

nagement needed at the different phases of a construction project. This would serve as a planning tool for a GMPCM to help in determining which of their management personnel need to be involved in the project, and at what time. The end result of that would be to give them more control of how the projects are to be managed.

CMAA's action words were studied to define the level of management responsibility and describe procedures in a management sequence. While this list does not supply specific processes, it does provide predefined words whose universal meaning makes it possible to create an understandable framework. These words also were used to determine if instructions within the framework contradict each other.

The proposed framework can be used to optimize the use of a PMIS when there is a disagreement between what is offered by the PMIS and procedure methodologies. It can also be used to restructure existing project procedure manuals to reconcile with the PMIS. This framework would provide a useable form of standardization to help with both this restructuring of existing procedure manuals and aid in the creation of new procedure manuals.

2. Current Use of Procedure Manuals

2.1. Preliminary Studies

A preliminary survey was conducted if contractors have enough written procedure manuals to effectively execute a construction project. Procedure manuals from Agency CM (ACM), Guaranteed Maximum Price CM (GMPCM), and owner-CM (OCM) contractors were mainly studied to see if there are significant differences among their management areas. The procedure manuals were categorized by management area for direct comparison. Another study showed large contractors' procedure manuals cover a wide range of management tasks from project administration, schedule, cost, contract administration, quality/safety to home office operations. In contrast, medium or small contractors' manuals focus on selected tasks such as site mobilization, startup, and closeout.

As part of the study of procedure manuals of contractors of various sizes the differences between a PMIS and the general contractor's procedures were studied. This study shows that PMISs were limited to collecting and distributing data, in other words a journal and accounting tool for a project. Procedure manuals were used to instruct project managers how management tasks should be executed. The trend usually though is that these manuals are used to supply the deliverable for each procedure.

2.2. Scope of Standard CM Services

Construction management is a professional management

service which applies to construction projects through controlling scope, time, cost, contract, and quality [4]. While there are some smaller builders who perform all of the work involved in a construction project the majority of time the main method involves the use of sub-contractors. Subcontracting is a primary delivery method to access an outside resource to perform a certain amount of work in the construction industry. The cost, quality and speed that come with working with subcontractors bring a few challenges. As each of these subcontractors is their own company they have their own reporting methods, document controls, and accounting. Without a well-organized management guideline, it is difficult for the constructor to effectively work with subcontractors due to the different management types of the subcontractor and the constructor.

CMAA's service map focuses on professional management practices in seven management areas such as project, cost, time, quality, contract administration, safety management, and program management. Nearly 200 tasks are defined from pre-design phase, design, procurement, construction, and post-construction phase [2]. Each management area contains a series of construction management tasks in five project delivery phases. A definition and level of responsibility for each task are provided.

General contractors have been implementing their own CM procedure manuals at the project level long before CMAA published the standard CM services in 1986. The CMAA's Construction Manager Certification Institute (CMCI), which administers the Certified Construction Manager program. This program is accredited by the American National Standards Institute and this standard was developed as part of CM certification. The standard contracts were developed to achieve the owner's goals [2], and serve as a way for owners and builders to communicate what the end result of a construction project will be.

It becomes an operational issue whether no integrated management system and procedures are available to implement or multiple systems and procedures are available. If there are none then there is the issue of having to create a new system and the costs and delays associated with that. With multiple systems and procedures there is the issue of interoperability between the systems and conflicts of the procedures, while there is no readily measurable costs there are the costs caused by delays, errors and the time spent in an effort to coordinate the systems. A revision of the systems is necessary to choose required services to meet the CM service requirements based on the size of the project organization structure, operation policies, owner's requirements, and project monitoring and control methods. Care should be taken to ensure that the proper system is chosen.

2.3. Organization of Procedure Manuals

Most CM procedure manuals focus on management processes and procedures in the construction phase, from project organization to project closeout. **Figure 1** shows a typical organization of the CM procedure manual. Large contractors tend to own a range of procedure manuals from site mobilization to project closeout. Normally the manuals are edited for each project. Based on the scope of project management, tasks, associated procedures, and particular forms are selected.

A task consists of at least three parts: one part explains what the intended purpose of the section is and why the procedure is important; another to assign the primary individual to implement the procedure; and the other to describe the steps to carry out the procedure. Often forms to fill out are provided to report the results of the procedure. Typically a set of manuals is edited based on the CM service requirements for the project before the construction phase. The procedure manuals are used more to support operational issues than professional management practices.

The participants include general contractors working in the U.S. and abroad: seven out of top 50 general contractors and four of them ranked in the top 50 contractors working abroad. According to the procedure manuals, about 330 reports are implemented by a general contractor on site. These reports are to cover all areas of the construction project, from finance/accounting, to business development, to industrial relations, to constriction support. Most of them (69%) are used to directly support construction operations such as controlling time, cost, or procurement. Quality control and reporting to assist field managers are also a focus. In addition, subcontract monitoring is one of the important components in construction support. 17 percent of the forms are related to

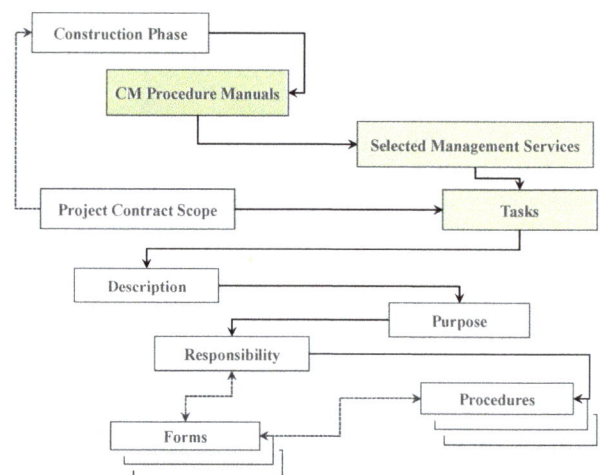

Figure 1. Structure of CM procedure manuals for construction phase.

finance/accounting and general accounting. The purpose of the forms is to monitor budget and expenses including payroll during the project period. Business development (3.6%) focuses on preconstruction support. Industry relations (10.6%) cover employment, public relations, and safety issues.

2.4. Use of Procedure Manuals

According to the survey, most procedure manuals are for site project operations. The scope of CM services is selected according to the management requirements of the request for proposal (RFP). In comparison the agreements created by the CMAA's describe the management areas and major tasks for two types of CM project delivery methods: Agency CM and GMPCM.

More than 36 procedures from 17 construction management (CM) companies were studied and the procedures are focused on site operation. Agency CM (ACM, 6%), guaranteed maximum price CM (GMPCM, 65%), and owner-CM (OCM, 29%) companies were included.

OCM procedures are larger than GMPCM's and ACM's procedures are typically the smallest. One of the OCM procedures includes nearly 40 management areas including typical owner tasks such as financing, engineering management, or owner supply items. GMPCM procedures cover 16 to 20 areas of project control at the corporate and site levels [5]. ACM procedures are the smallest in term of service areas and cover management services of contractually requested. Thus it is hard to describe the character of ACM procedures.

Table 1 shows the service modules described by the selected contractors. Of note, larger contractors tend to have an integrated procedure to provide seamless construction management services. These included project administration, time, cost, contract administration, quality, and safety. The three cost related service areas such as scope, time, cost, and contract management are crucial since they focus on controlling and monitoring of schedule/cost. These contractors have the resources to create these seamless systems, and also tend to focus on schedule and cost controls. Larger contractors also had extensive document control procedures as part of their manuals.

For the medium and small contractors, site operation was a primary concern. Their manuals tend to have sectional procedures of specific tasks such as pre-construction, mobilization, project closeout, public relations, or commissioning. A schedule/cost module was also included as a core module. However, procurement or contracting was included but was not fully integrated with the schedule/cost module. Base modules such as project administration policy or project numbering systems were used by larger contractors.

Table 1. Service areas covered by selected procedures.

Service Areas	Contractors			
	A	B	C	D
Management & Administration	√	√	√	√
Organization & Responsibility	√	√	√	√
Project Control	√	√	√	√
Scope Management	√	√	√	
Time Management	√	√	√	
Cost/Estimating	√	√	√	√
Safety Management	√	√		
Quality Control/Inspection	√	√	√	
Contract Administration/Procurement	√	√	√	√
Project Numbering System	√	√		
Project Management System				
Document Control System	√	√		√
Progress Reports	√		√	√
Change Management	√		√	√
Delays & Claims	√			√
Quality Assurance			√	
Quality Plan			√	
Request for Information	√			

Design management procedures were occasionally found since it is not a required task in building construction projects. Design control, drawings and specifications, design review were only found in industrial or heavy/civil projects.

Necessary but special services such as constructability review, risk analysis, value engineering, and life-cycle-cost analysis were developed as additional procedures since they require design engineering and construction engineering knowledge.

2.5. Current Practices in Computerization

A PMIS is an integrated information system originally intended to assist project managers to collect, analyze, or distribute project information [6]. Generally a PMIS contains predefined business processes and procedures, to aid in data collection and dissemination Lee [7] verified that the chance to complete a project successfully can be increased by 75% if the right PMIS is implemented.

Three approaches are generally used for computerizing the CM business process to create a PMIS: commercial PMIS, the "off the shelf" products (51%), custom-built

PMIS, literally created for the needs of the construction company (18%), and the hybrid PMIS, a commercial system with some custom built modules added on (31%). While only a portion of the contractors have an entirely custom-built PMIS, in order to keep the consistency of their internal processes and procedures, a significant number of them would prefer a custom-built PMIS.

No evidence suggested these services are fully computerized. Some contractors implemented a custom built PMIS which was developed based on the contractors' procedures. Their forms and procedures are fully embedded in the PMIS. For contractors who use a comercial "off the shelf" PMIS often use their own forms because they do not want to give up their existing management procedures. In essence they have a computerized system while they continue to fill out their own system of papers. Because these forms are not fully embedded in this type of PMIS this makes some contractors hesitant to implement a commercial PMIS. When it is used, the PMIS is often being used as a data collection, analysis or distribution tool. What is not being used is the ability of a PMIS to fill out multiple types of forms with one set of data. Time is being spent adding the same data multiple times into different forms used by the contractor.

Another issue, due to the expense, custom PMISs have frequently not been updated as the procedure changes or a new project management method is introduced. This widens the difference between the procedures and a PMIS. As a result, most contractors rely on off-line management with the PMIS being used more as a data management tool instead of a project management tool. This negates the effectiveness of the PMIS and discourages the implementation of the PMIS at the project level.

3. Proposed Framework of Procedure Manuals

To propose a framework which integrates management modules, studies if standard bidding practices, project management and their functions, management tasks. The management areas were divided into two groups: project control modules (scheduling, budgeting, and contracting) and other supporting modules. Action words and management tasks were studied to propose a framework not only for integrated management but also for business process management of PMIS.

3.1. Proposed Framework

The proposed framework for the integrated management of time, cost, and contract is illustrated in **Figure 2**. By defining the scope of a project, schedule data, budget data, and procurement/contracting data can be fully integrated. The framework calculates progress measurement

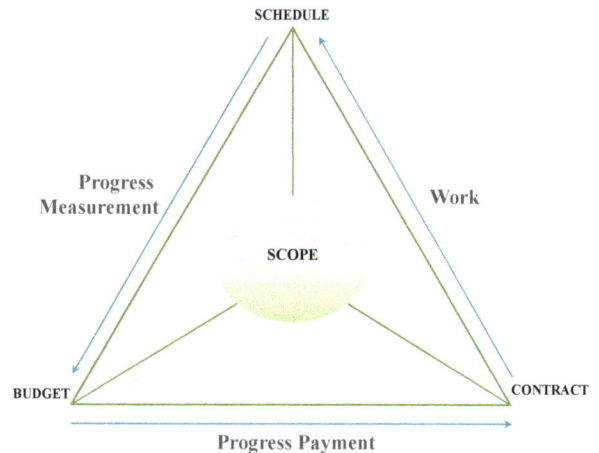

Figure 2. Core management services.

and progress payment easily. This framework also allows managers to update or revised the scope of project when a change order is considered.

Figure 3 maps out how the CM services should support each other at various stages of construction business operations. When the RFP is announced, general contractors submit a bid package in order to be considered for the project. A bid package includes all CM plans which are produced based on procedure manuals with selected management modules. Eventually, work process frameworks can be concentrated into the agreement phase when the general contractor's submittal is reviewed and meets the owner's requirements. Management information can be, simultaneously, collected, organized, and transmitted to the owner using the PMIS.

Figure 4 shows the minimum four levels recommended to combine project management information into one pool of knowledge.

At Level 1, information of the project progress, the project cash flow, and the contract changes is critical. Three pieces of information should be accessed through an activity. Thus not only all deliverables of the project but also the control level activities should be identified. By defining the project deliverables (scope), the remaining management components are connected through the relationships between change orders, progress measurement, and progress payments. When a change order is approved, the scope will be revised and the schedule can be revised. A progress payment can also be made when a progress is measured.

At Level 2, any information affecting scheduling, budget, and contracting should be considered. Creating a project work breakdown structure (PWBS) is critical; the scope of the project is identified. Resource information should be attached so that cost related analysis can be completed. Measuring the time or cost impact caused by a change or delay becomes possible. This also allows

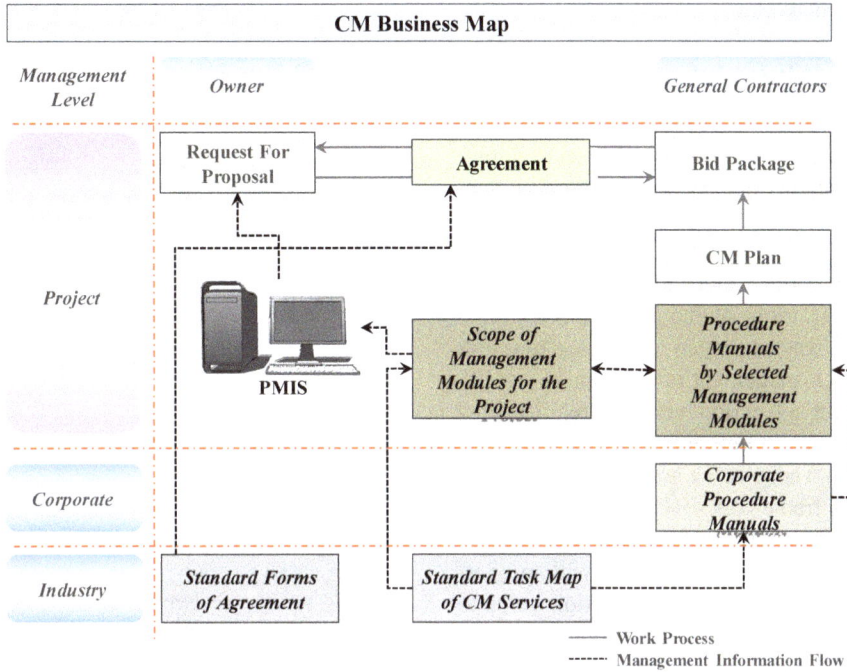

Figure 3. Proposed management framework (Revised from [2]).

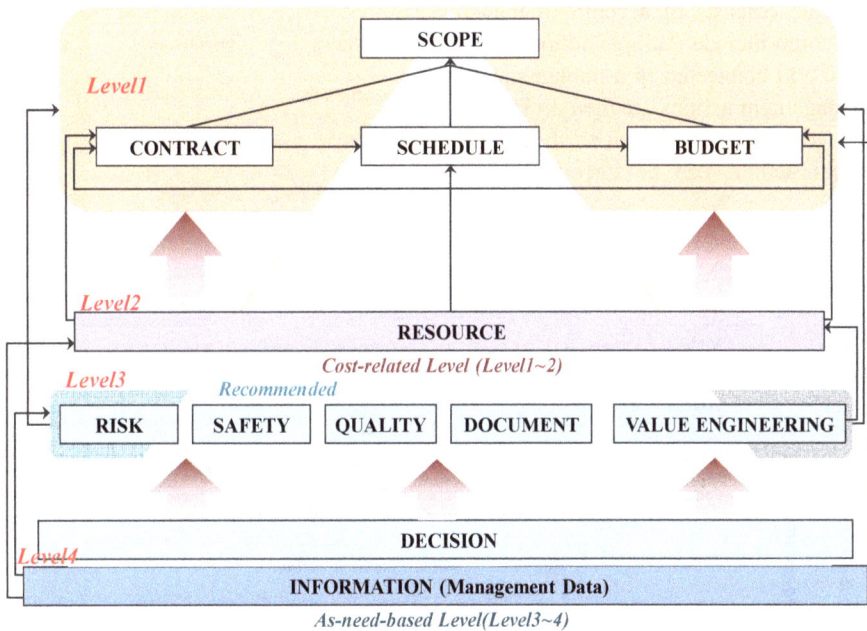

Figure 4. Hierarchy of key management modules.

resource allocation and planning at the corporate level. This can be implemented relatively easily because most large contractors implement the earned value method (EVM) for this purpose.

At Level 3, non-cost information such as safety or quality can be attached. Technical modules such as risk analysis, value engineering can also be included in this level. These modules are supporting modules. Thus they

can be developed and integrated independently since they produce information that is needed for limited decision making. Owner requested service, e.g., value engineering or constructability review can be added to this level.

Level 4 contains the decision making module which illustrates how project information is used to make a management decision. Due to a difference in management strategies or operation policies of a contractor, it requires a custom framework.

3.2. Tasks and Procedures Integration Map

To provide quality CM services, at least twelve management modules are suggested; budget, project administration, contract, schedule, quality, safety, resource, material/equipment, decision, risk, information management, and value engineering [8]. As Uren [9] suggested ontologies, documents, annotations that link the ontologies to the document, and the users of the systems are implemented. This method was employed to increase reusability and configuration management of the framework [8]. As in **Table 2**, about 63 action verbs were identified from the CM agreements to define the level of responsibility in construction. These verbs are the core elements of the ontology of the framework.

Figure 5 shows an ontology map for schedule management, monitoring schedule compliance, with proposed tasks and procedures. The map shows action verbs to describe the management needs, management goals, and project data required for the task. In order to remove any ambiguity, each action verb must be defined. Thus the effects of each word can be traced.

A management module consists of a chain of management tasks. A task could include multiple actions. An action word is selected and connected to a management action. A series of management actions are then arranged by chronological management order of each as shown in **Figure 5**. Management actions can be traced or observed.

Table 2. List of action words.

address	develop	investigate	propose
administer	discover	issue	provide
apprise	discuss	maintain	quantify
approve	distribute	make	recommend
arrange	document	manage	record
assemble	establish	measure	reduce
assess	estimate	meet	report
assist	evaluate	modify	review
assure	expedite	monitor	revise
attend	explain	negotiate	require
author	facilitate	notify	summarize
chair	follow-up	open	tabulate
clarify	formulate	organize	transmit
compare	forward	pay	update
conduct	identify	perform	verify
contact	implement	plan	
coordinate	incorporate	prepare	
define	initiate	price	
determine	interview	produce	

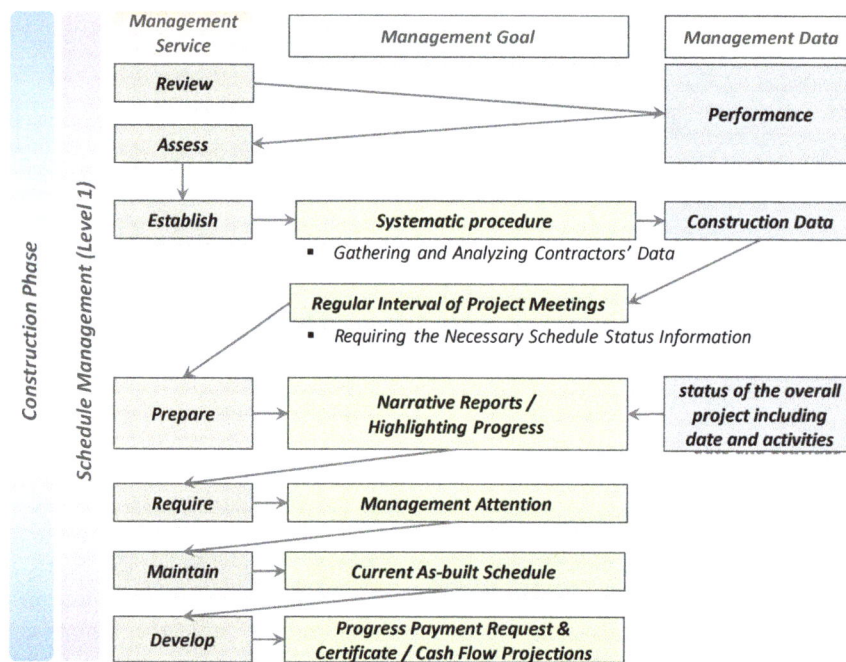

Figure 5. Sample framework of the proposed tasks and procedures integration map.

It is critical to use the same action words in the procedures and PMIS to remove any misinterpretation. They also explain the level of obligation of each task. A chain of actions words are then used to define the accountability of each responsible party.

3.3. Suggestions in PMIS Development

To effectively develop a PMIS, business process modeling is required. It requires a study of current business processes and how the procedures will be reengineered to improve effectiveness. After the reengineering has taken place or if the procedure does not need improvement then the procedures can be used to select a PMIS that fits the management requirements if a commercially available PMIS is considered.

This first part of this procedure is used in the creation of a custom built PMIS. The reengineered procedures are then used by the programmer to create the custom made system to match up with the procedures and reports. For a hybrid system, if the commercial system does not have anything to match up with some of the procedures and the procedures are deemed necessary then customized modules are created to address those procedures.

By following a standard software development methodology, management processes and procedures will be mapped out. Interactions between PMIS and other information systems such as the enterprise resource planning (ERP) system, supply chain management (SCM) are commonly considered. The industry specific software including scheduling and estimating is also considered. This can save time in the development phase since the procedures, especially forms, are what describe document transactions between systems. Thus integration among management systems can be achieved easily and in a manner that makes implementation at the project level more effective due to the system method matching the project level interactions.

4. Conclusion and Future Work

Based on studies of more than 36 procedures manuals from construction management companies, the following observations are summarized:

Full service modules are implemented by large contractors whereas medium or small contractors implement a limited number of modules solely for site operations. These are: schedule/cost-oriented, document control-oriented, and site operation-oriented.

Large contractors focus on the controlling and monitoring of the entire project. This is to assist construction managers at sites as well as home offices. A schedule/cost-oriented framework was used to measure the performance of the project using a schedule and cost information. Extensive project operation procedures are often

used to cover a wide range of particular CM services such as VE, LCC, CR, risk analysis, or public relations.

Level of integration of the procedures with comercial PMIS is relatively low. The PMIS is mainly used as a central data repository for project control or document control. Off-line construction operations are still handled by separate procedures. This drops the efficiency of PMIS implementation. This makes contractors hesitant about purchasing a commercial PMIS package. Finding a PMIS package that fits the corporation's project control procedures is a difficult process.

Medium or small contractors tend to implement modules for site operations at the project site such as preconstruction, mobilization, commissioning, and project closeout. Because of a lack of technical analysis, trending and forecasting are challenging. Formulating procedures in the project control areas are necessary and computer software should be employed to cover missing project control methods.

Even though the PMIS is an effective data processing tool in project management the synchronization of off-line operations and on-line reporting procedures is recommended in order to improve the efficiency. As part of the procedure manual the functionality requirements of the PMIS should be described. Thus selection of a PMIS with synchronized management modules is important.

REFERENCES

[1] Y. Lin and H. Lee, "Developing Project Communities of Practice-Based Knowledge Management System in Construction," *Automation in Construction*, Vol. 22, 2012, pp. 422-432.

[2] CMAA, "Construction Management Standards of Practice," Construction Management Association of America, 2010.

[3] C. Thomsen, "Program Management Version 2.0," Construction Management Association of America, 2008.

[4] CMAA, "An Owner's Guide to Construction and Program Management: Enabling Project Success under Any Delivery Method," Construction Management Association of America, 2011.

[5] B. Y. Ryoo and J. Kang, "Impact of CM Procedure Manuals in Construction Project Management," *3rd International Conference on Construction in Developing Countries*, Bangkok, 4-6 July 2012.

[6] A. Jaafari, "Towards a Smart Project Management Information System," *International Journal of Project Management*, Vol. 16, No. 4, 1998, pp. 249-265.

[7] S. Lee and J. Yu, "Success Model of Project Management Information System in Construction," *Automation in Construction*, Vol. 25, No. 10, 2012, pp. 82-93.

[8] C. Haltenhoff, "The CM Contracting System: Fundamentals and Practices," Prentice Hall, Upper Saddle River,

1998.

[9]   V. Uren, "Semantic Annotation for Knowledge Manage-
      ment: Requirements and a Survey of the State of the Art,"
      Web Semantics: *Science, Services and Agents on the
      World Wide Web*, Vol. 4, No. 1, 2006, pp. 14-28.

A Comparative Analysis of Job Stress of Field Managers and Workers in Korean Construction Projects

Zhen Zhang, Woo-Hwan Lee, Young-Wha Choi, Sung-Hoon An[*]

Department of Architectural Engineering, Daegu University, Gyeongsan-Si, South Korea.

ABSTRACT

A successful construction project hinges on the effective and efficient management of human resources. The stress of human resources is directly related with work performance, and as such, should be managed to improve work performance. This study aims to perform a comparative analysis of job stress levels after a survey on the job stress of field managers and workers. Through the analysis, it is found that the stress levels of field managers are different from those of the workers and construction field managers and workers get job stress scores less than average scores of Korean male workers. In addition, different personal factors affect the field managers and the workers differently. Therefore, understanding by which factor the other parties become stressed is expected to improve efficiency in the management of human resources.

Keywords: Job Stress; Comparison; Field Manager; Worker; Korea

1. Introduction

Every construction project involves the participation of many different people. For this reason, the success of a construction project relies heavily on how efficiently and effectively the human resources (manpower) are managed [1,2]. For the successful completion of a construction project, the human resources must be provided with an environment in which they can engage in their tasks faithfully [3]. Stress can have either positive or negative effects. That is, if the stress level is appropriate, it will help give the human resources a positive tension, improving their work efficiency as a result, while if the stress level is too high, it will have an adverse influence on the work efficiency and capability of the human resources [4,5]. In particular, as the stress level of the human resources has a close relation with safety accidents, it should be managed in order to reduce safety accidents [6,7].

Many studies have been conducted on stress management for human resources [4-14]. The previous studies were conducted either on field managers or on individual workers. However, the stress level of field managers can be different from that of workers, even in the same construction site. If the stress influencing factors are understood by the participants (field managers or worker), it is

expected that human resources will be effectively managed, and the stress level can be lessened through efforts to relieve the other's stress.

Therefore, the goal of this study is to perform a comparative analysis after a survey of the job stress of field managers and workers in Korean construction sites.

2. Job Stress

It is difficult to find a clear-cut definition of the term "job stress" in the literature. Sun [15] defined job stress as "a physical and emotional phenomenon experienced in various working situations, when workers are at risk, when they have a conflict with their boss or colleagues, when they feel emotional affliction arising out of a discrepancy between their aptitude and their job, when their ability is not recognized or they feel they lack ability, or when they are in charge of an important or extremely difficult job." In short, one can experience job stress when one is not given a working environment in which one can control the job, or with which one's motivation or capability is compatible.

If one is under job stress continuously and the stress has been accumulated without being released, one can experience a deterioration in one's physical condition or family discord, frequent absences from work or frequent job changes, high emotional disturbance and maladjustment, and low performance [11,12]. Therefore, job stress is one of the main factors that have an influence on the organizational performance.

3. Research Method

The scope of this research was limited to field managers and workers working on Korean construction sites of the top 30 Korea construction companies, because the job stress of human resources can be different depending on the company size (large or small). More specifically, the working environment (including working condition and remuneration), which are depending on the company size, can affect stress level differently. In addition, the subjects studied in this research, both field managers and workers, were limited to men only, since the field managers and workers in the construction site are usually men.

To compare job stress between field managers and workers, the method of measuring job stress was first selected. Job stress was measured through questionnaires. The questions on the questionnaire were selected and considered to be most appropriate for this study after reviewing the questions used in the previous studies. The data was collected from February through May, 2013.

The job stress of field managers and workers can be measured through the stress influencing factors. There are many research papers that mention different methods of measuring job stress through stress influencing factors [4-14]. We used the "Occupational Stress Scale for Korean Employees" developed by the Korea Occupational Safety and Health Agency (KOSHA) to measure the occupational stress of the field managers of construction projects [16], because the "Occupational Stress Scale for Korean Employees" was developed by the research team of KOSHA after reviewing many occupational stress instruments, it is appropriate for Korean workers, and its validity has been verified through many cases.

There are two versions of the "Occupational Stress Scale for Korean Employees": the basic version (43 questions) and the shortened version (24 questions). The shortened version with 24 questions was employed in this study, since the basic version has so many questions that it can be difficult for respondents to complete.

The shortened version consists of 24 questions in 7 categories (job demand, job control, interpersonal conflict, job instability, organizational system, inappropriate compensation, and organizational climate). The details of 24 questions are expressed in **Table 1** [16].

4. Comparative Analysis of Job Stress

4.1. Reliability and Validity of Variables

To measure the job stress in 7 categories, the questionnaire was comprised of different questions. Therefore, the reliability must be analyzed using Chronbach's Alpha to verify the consistency of the questions in each category, and to perform a factor analysis to determine whether the questions are appropriate.

However, the reliability and validity were not verified in this study, because the occupational stress scale used in this study had already been verified in a previous study [16]. In addition, the questions should be identical to compare the job stress measured in this study with Korea standard occupational stress.

4.2. A Comparative Analysis of Job Stress

The job stress of field managers and workers in construction projects was measured through a survey. As mentioned earlier, the abridged version of the "Occupational Stress Scale for Korean Employees" developed by KOSHA was used in this study. The survey results were scored by category using Equation (1) [16]. If the job stress scores are lower, the workers are less stressed.

$$\frac{\left(\text{sum of score-number of question}\right)}{\left(\text{highest score-number of question}\right)} \times 100 \qquad (1)$$

The job stress scores of field managers and workers shown in **Table 2** were compared with the average job stress scores of Korean men announced by KOSHA. As indicated in **Table 3**, the job stress score of field managers was shown to be considerably lower than the average, while the job stress score of workers was shown to be slightly lower than the average.

Table 3 shows the results of a t-test to statistically verify the job stress scores between the field managers and workers. Through the t-test results, it was revealed that work stress scores were statistically different in all of the 7 categories, at a 95% confidence level ($p < 0.05$).

As shown in **Table 2**, the job stress scores of field managers are remarkably different from those of workers, which showed an entirely different tendency. Therefore, it is revealed that the factors that have an influence on the job stress of field managers are believed to be different from those that affect the worker's stress, and it is verified by T-test as shown in **Table 3**.

4.3. Review of the Comparison Results of Job Stress

In terms of job demand, the score of the field mangers was similar to the average, while the score of the workers was considerably lower than the average. This may be because the construction workers do not receive much stress from the workload they carry and the rest they take.

In terms of job control, the score of field managers was shown to be considerably lower, while that of workers

Table 1. Details of the shortened version with 24 questions.

Categories	Questions
Job demand	I am always under pressure to meet my deadlines due to the heavy workload. The workload has been significantly increased. I can get sufficient rest during work. I have to deal with a couple of jobs at the same time.
Job control	I need to be creative to perform my job. I need a high level of skill and knowledge to perform my job. I am empowered to determine the working time and job performance, and I can make a decision at my discretion. I can determine my workload and change my work schedule at my discretion.
Interpersonal conflict	I can get some help from my boss to finish my work. I can get some help from my colleagues to finish my work. There are people at work who can understand me and my situation whenever I have difficulty with my job.
Job instability	Our company is unstable, and the future of my job is insecure. There was an undesirable change in my working condition (e.g. layoff) or I expect such a change.
Organizational system	Our company has a fair and reasonable system for work performance appraisal and personnel (promotion, department placement, etc.). Our company has a good supporting system including staff, space, facility and training if necessary. Our company has an efficient system that enables cooperation between departments without conflict. I can find an opportunity and path to reflect my ideas.
Inappropriate compensation	Considering my efforts and performance, my remuneration is appropriate. Considering that my situation is expected to become more favorable, I do my job without feeling burdened. I am given opportunities to develop and show my ability.
Organizational climate	I feel uneasy at office dinners. I am given work instructions with no consistent criteria. The working atmosphere is authoritative and vertical. I get discriminated against sexually.

Table 2. Comparison of job stress with Korean workers' standard.

Element	Korean workers' standard (A)	Field managers (B)	Workers (C)	B − A	C − A	B − C
Job demand	52.40	54.38	46.12	1.98	−6.28	8.26
Job control	53.70	38.59	57.19	−15.11	3.49	−18.6
Interpersonal conflict	41.20	32.07	40.19	−9.13	−1.01	−8.12
Job instability	50.13	40.53	46.75	−9.6	−3.38	−6.22
Organizational system	52.78	43.33	49.11	−9.45	−3.67	−5.78
Inappropriate compensation	52.11	41.75	48.96	−10.36	−3.15	−7.21
Organizational climate	40.95	35.44	42.48	−5.51	1.53	−7.04
Average	49.03	40.85	47.24	−8.18	−1.79	−6.39

Table 3. Result of T-test for job stress.

Type	Field managers (N = 99)		Workers (N = 96)		T-value	p-value
	Mean	S.d	Mean	S.d		
Job demand	54.38	16.622	46.12	9.985	4.222	0.000
Job control	38.59	14.007	57.19	12.163	−9.884	0.000
Interpersonal conflict	32.07	15.400	40.19	14.463	−3.795	0.000
Job instability	40.53	20.678	46.75	15.416	−2.384	0.018
Organizational system	43.33	12.447	49.11	8.366	−3.816	0.000
Inappropriate compensation	41.75	13.550	48.96	12.520	−3.853	0.000
Organizational climate	35.44	15.106	42.48	10.449	−3.794	0.000
Average	40.85	8.975	47.24	5.867	−5.896	0.000

was all shown to be slightly higher than the average. For the field mangers have high authority, and they can make the decision on working time and workload in their discretion; for this reason, they became less stressed.

In terms of interpersonal conflict, job instability, organizational system and inappropriate compensation, the scores of field managers and workers were shown to be lower than the average. In particular, the field managers were shown to become remarkably less stressed. It is supposed that the field managers surveyed in this study are the employees in the top 30 construction companies, and they are overall satisfied with their company.

In terms of organizational climate, the score of field managers was shown to be lower than the average. This is believed that people in an organization at construction site have a sense of friendship with each other.

4.4. Job Stress Influencing Factors

Based on the differences between field managers and workers in job stress scores, the job stress influencing factors were expected to be different. Therefore, to understand the respective job stress factors influencing either the field managers or the workers, as shown in **Tables 4** and **5**, t-test was conducted by marriage, hobby, and exercise and ANOVA was conducted by ageand character.

In terms of personal factors that have an impact on the job stress of the workers, as shown in **Table 4**, the older the field managers and workers are, the less stressed they become. In terms of job instability, the workers in their 20s and 30s were less stressed compared to those in their 40s and 50s. In terms of job demand, job instability and organizational system, the married workers were more stressed than those who were still single. In terms of inappropriate compensation, the workers who had a hobby were less stressed than those who did not have a hobby. In terms of job control, the workers who exercised regularly were less stressed than those who did not exercise regularly. Character was found to have an influence on the stress related with job demand and job control.

In terms of personal factors that can influence the job stress of field managers as shown in **Table 5**, the older the managers were, the more stressed they were regarding job instability. Married field managers were more stressed regarding job demand while less stressed regarding job control compared to the unmarried. In terms of inappropriate compensation, field managers with a hobby were less stressed than those who did not have a hobby. In terms of interpersonal conflict, the field managers who exercised regularly were less stressed than those who did not exercise regularly. Character was not shown to have an influence on the stress.

Through the analysis of the influence on stress levels of site managers and workers by personal factor (age, marriage, hobby, exercise and character), it was found that some personal factors had an impact on a specific stress. However, each personal factor had a different impact on the stress level of the workers from that of the field managers.

5. Conclusions

In this study, the job stress of field managers and workers in construction projects was measured. The results showed that the job stress of field managers was considerably lower than the average job stress of Korean men, while the job stress of the workers was slightly lower than the average. In addition, in the 7 categories of job stress, there were remarkable differences shown between the field managers and the workers. In particular, in terms of job control, the field managers became outstandingly less stressed compared with normal Korean men, because the field managers had high level of autonomy due to the nature of a construction project, and they can make a decision regarding working time and workload, which makes them less stressed. In addition, in terms of interpersonal conflict, job instability, organizational system, and inappropriate compensation, the scores of field managers were much lower than the average. One of these reasons is supposed that field managers are satisfied with their companies because they are employed by top 30 construction companies of Korea.

Through the analysis of personal factors (age, marriage, hobby, exercise, and character), a specific personal factor was found to have an impact on a particular category of job stress. However, the personal factors were found to differently affect the stress levels of field managers and the workers.

It is far-fetched to say that these research findings would bring the same results from all field managers and workers in construction projects because the field managers and the workers were limited to those employed by Korea's large construction companies. In addition, it is difficult to generalize the analysis results of the human resources. Therefore, it is necessary to study field managers and workers who work for small- and medium-sized construction companies in the future to determine whether these research findings are compatible.

If field managers and workers can understand the factors that cause stress to the other party, the human resources can be managed more effectively. This is expected to enable efficient management of the organization at construction sites and raise work performance.

6. Acknowledgements

This research was supported by Basic Science Research Program through the National Research Foundation of

Table 4. Factors to effect the workers' job stress (N = 96).

Factor		N	Job demand	Job control	Interpersonal conflict	Job instability	Organizational system	Inappropriate compensation	Organizational climate	Average
Age	~29	4	45.35	67.20	50.00	40.63	54.70	60.40	43.80	51.73
	30 - 39	17	41.93	62.90	40.68	36.77	44.14	52.46	41.21	45.69
	40 - 49	40	47.84	56.74	41.26	48.44	48.78	46.87	44.09	47.70
	50~	35	46.27	53.77	37.62	50.36	51.28	48.33	41.10	46.94
	p-value		0.241	0.023*	0.369	0.015*	0.015*	0.117	0.606	0.280
Marriage	Yes	82	47.43	56.66	40.65	48.93	49.87	48.47	42.48	47.77
	No	14	38.41	60.27	37.49	33.93	44.66	51.79	42.44	44.13
	p-value		0.001*	0.305	0.604	0.001*	0.030*	0.362	0.989	0.164
Hobby	Yes	80	45.89	56.98	40.42	46.25	49.32	47.60	42.53	46.98
	No	16	47.29	58.23	39.06	49.22	48.08	55.73	42.22	48.53
	p-value		0.611	0.708	0.734	0.485	0.589	0.017*	0.914	0.337
Exercise	Yes	63	47.25	54.89	39.55	44.84	48.34	47.22	43.68	46.52
	No	33	43.96	61.57	41.42	50.38	50.59	52.27	40.18	48.61
	p-value		0.088	0.004*	0.506	0.064	0.212	0.060	0.120	0.053
Character	leading	8	32.05	68.76	38.55	40.63	42.98	56.25	45.34	46.36
	prudent	36	46.03	55.41	39.82	44.79	50.20	46.75	42.04	46.41
	stable	42	48.69	56.42	40.67	47.92	48.98	49.41	43.19	47.88
	social	10	46.89	57.53	40.83	53.75	50.66	49.17	38.77	48.21
	p-value		0.000*	0.039*	0.979	0.248	0.153	0.276	0.556	0.650

*Gap is significant at the 0.05 level (2-tailed).

Table 5. Factors to effect the field managers' job stress (N = 99).

Factor		N	Job demand	Job control	Interpersonal conflict	Job instability	Organizational system	Inappropriate compensation	Organizational climate	Average
Age	~29	5	36.28	37.52	21.66	20.00	32.50	31.66	18.76	28.32
	30 - 39	62	56.48	40.04	33.47	37.90	43.17	43.01	35.71	41.38
	40 - 49	30	53.36	36.69	31.94	47.92	45.65	41.12	37.32	41.98
	50~	2	50.00	25.05	16.65	62.50	40.65	37.50	40.65	39.00
	p-value		0.064	0.379	0.187	0.006*	0.177	0.315	0.076	0.013*
Marriage	Yes	63	57.37	35.74	32.80	42.86	43.58	40.88	34.74	41.12
	No	36	49.16	43.59	30.78	36.46	42.90	43.28	36.65	40.39
	p-value		0.017*	0.007*	0.533	0.139	0.794	0.398	0.548	0.697
Hobby	Yes	83	55.00	37.52	31.02	40.21	42.72	40.56	35.42	40.33
	No	16	51.19	44.16	37.50	42.19	46.51	47.93	35.56	43.56
	p-value		0.405	0.083	0.124	0.728	0.268	0.046*	0.973	0.190
Exercise	Yes	66	53.91	37.90	29.16	37.69	42.74	39.90	34.58	39.39
	No	33	55.33	39.97	37.88	46.21	44.53	45.45	37.15	43.77
	p-value		0.690	0.491	0.015*	0.053	0.501	0.054	0.428	0.021*
Character	leading	9	50.02	31.27	37.96	40.28	44.47	40.74	30.58	39.30
	prudent	36	57.14	40.47	33.09	42.71	41.84	41.90	38.04	42.16
	stable	38	52.00	38.18	29.38	38.82	43.12	41.89	35.22	39.78
	social	16	56.28	39.47	32.82	39.84	46.51	41.66	32.84	41.33
	p-value		0.463	0.365	0.451	0.882	0.659	0.996	0.485	0.661

*Gap is significant at the 0.05 level (2-tailed).

Korea (NRF) funded by the Ministry of Education, Science and Technology (2011-0021835).

REFERENCES

[1] S.-H. An, "A Study on the Construction Manager's Leadership Styles Based on Condition of Building Projects," *Journal of the Architectural Institute of Korea*, Vol. 25, No. 4, 2009, pp. 231-238.

[2] D. Lee, M. Kim and S. Kim, "Management Performance Evaluation Model of Korean Construction Firms," *Journal of Building Construction and Planning Research*, Vol. 1, No. 2, 2013, pp. 27-38.

[3] N. Inyang, M. Al-Hussein, M. El-Rich and S. Al-Jibouri, "Ergonomic Analysis and the Need for Its Integration for Planning and Assessing Construction Tasks," *Journal of Construction Engineering and Management*, Vol. 138, No. 12, 2012, pp. 1370-1376.

[4] M.-Y. Leung, I. Y. S. Chan and P. Olomolaiye, "Impact of Stress on the Performance of Construction Project Managers," *Journal of Construction Engineering and Management*, Vol. 134, No. 8, 2008, pp. 644-652.

[5] M.-Y. Leung, I. Y. S. Chan and C. Dongyu, "Structural Linear Relationships between Job Stress, Burnout, Physiological Stress, and Performance of Construction Project Managers," *Engineering, Construction and Architectural Management*, Vol. 18, No. 3, 2011, pp. 312-328.

[6] M.-Y. Leung, I. Y. S. Chan and J. Yu, "Preventing Construction Worker Injury Incidents through the Management of Personal Stress and Organizational Stressors," *Accident Analysis & Prevention*, Vol. 48, 2012, pp. 156-166.

[7] O. O. Abbe, C. M. Harvey, L. H. Ikuma and F. Aghazadeh, "Modeling the Relationship between Occupational Stressors, Psychosocial/Physical Symptoms and Injuries in the Construction Industry," *International Journal on Industrial Ergonomics*, Vol. 41, No. 2, 2011, pp. 106-117.

[8] H.-J. Choi and H.-G. Kwon, "The Stress Influences on the Job Attitudes toward the Employees of Construction Industries," *Korean Journal of Business Administration*, Vol. 21, No. 4, 2008, pp. 1723-1749.

[9] J.-H. Eom, "A Study on the Stress Control of Construction Workers," Master Thesis, Seoul National University of Technology, Seoul, 2003.

[10] M.-Y. Leung, I. Y. S. Chan and J. Yu, "Integrated Model for the Stressors and Stresses of Construction Project Managers in Hong Kong," *Journal of Construction Engineering and Management*, Vol. 135, No. 2, 2009, pp. 126-134.

[11] P. Bowen, P. Edwards and H. Lingard, "Workplace Stress Experienced by Construction Professionals in South Africa," *Journal of Construction Engineering and Management*, Vol. 139, No. 4, 2013, pp. 393-403.

[12] P. E. D. Love, D. J. Edwards and Z. Irani, "Work Stress, Support, and Mental Health in Construction," *Journal of Construction Engineering and Management*, Vol. 136, No. 6, 2010, pp. 650-658.

[13] S.-H. An, Z. Zhang and U.-K. Lee, "Correlation between Job Stress and Job Satisfactionof Building Construction Field Managers," *Journal of the Korea Institute of Building Construction*, in Press, 2013.

[14] T.-H. Jeong, "The Study of an Analysis on Job Stress of Affecting Safety and Management Performance for Workers in the Construction Industry," Ph.D. Thesis, Chosun University, Gwangju, 2009.

[15] J.-W. Sun, B.-S. Oh, D.-S. Hwang and J.-Y. Kim, "An Introduction to Job Stress," Korean Studies Information, Paju, 2010.

[16] J.-W. Sun, B.-S. Oh, D.-S. Hwang and J.-Y. Kim, "A Measurement of Job Stress," Korean Studies Information, Paju, 2010.

9

Waste Shell Husks Concrete: Durability, Permeability and Mechanical Properties

Md. Zakaria Hossain

Department of Environmental Science and Technology, Graduate School of Bioresources, Mie University, Tsu, Japan.

ABSTRACT

Shell husk is annually produced as a byproduct of shell production in Japan. According to Japanese Ministry of Forestry, Fisheries and Agriculture, the amount of the abandoned shell husk is about 151,000 tons per year. This huge amount of abandoned shell husk is not only thrown away without any commercial return but also causing pollution and environmental problems. To mitigate the pollution and environmental problems, possible utilization of abandoned shell husk is thoroughly observed in concrete construction. Overall response of the mechanical properties of concrete specimens containing different percentage of abandoned shell husk aggregates such as 0, 10%, 0, 30%, 40% and 50% in the ratio of mass is demonstrated. Results of engineering properties such as compressive strength, Young's modulus, tensile strength, unit weight, water absorption capacity and coefficient of hydraulic conductivity are depicted. It is observed that the use of shell husk in concrete improves strength and durability performance of concrete treated in aggressive sea environments.

Keywords: Shell Husk; Recycle; Concrete; Durability; Mechanical Properties; Permeability

1. Introduction

A huge amount of abandoned Mactridae shell husk is produced in Japan every year. Mactridae, also known as trough shells or duck clams, is a family of marine bivalve clams of the order Veneroida. In Japan, the amount of the abandoned shell husk is about 151,000 tons per year according to the Japanese Ministry of Forestry, Fisheries and Agriculture. Among these, about 9627 tons/year are produced in Mie prefecture and 715 tons/year are produced in Tsu city along with 407 tons/year of mactrachinensis (mactridaes) in Tsu city. According to the information of disposal site office in Tsu city, the cost for disposing the abandoned shell husk is about 18 yen per kilogram which means that nearly 151411US$ is required for disposing abandoned shell husk in Tsu city (nearly 86188US$ is required for mactridaes) only. In addition to this, nearly 2 million US$ is required in Mie Prefecture and nearly 32 million US$ is required in Japan. These mactridaes, which are not only thrown away without any commercial return and a lot of money is being spent for its disposal but also causing pollution and environmental problems. With the rapid increase in demand for balance between natural phenomena and ecol-

ogy in bio-environment, there is a significant need for continuous development of new technologies that consume these abandoned shell husks [1,2]. One idea is gaining much attention, lately, is that of the use of abandoned mactridae shell husk as recycled aggregates for the manufacture of concrete or cement-based composite materials. This solves mainly two environmental and economic aspects, such as: 1) solving the waste storage problem and 2) protection of limited natural resources of aggregates [3,4]. Both the economic and environmental benefits can be achieved by using the abandoned shell husk as recycled aggregate in concrete or in cement-based composites. Some researches on concrete or cementitious composites with other recycled aggregate produced from building destruction can be found in the technical literature [5-8]. Light weight concrete masonry with recycled wood aggregate was studied by Stahl [9]. The latest researches on the assessment of the surface resistance and permeation properties of recycled aggregates concrete are also found [10,11]. In spite of the volume of available technical information, there is a little research work on the use of abandoned shell aggregates in concrete or cement-based composites [12]. In view of the above distinct advantages for environmental conservation and protection of limited natural resources as well as economy; the present research work is undertaken to fulfill this basic need. It is expected that the outcomes of this research assist in possible utilization of abandoned shell husk as aggregate for producing a suitable concrete or cement-based composite for prospective uses in the field such as; terrace lands, roads, canals, pavements, embankments, ridge between paddy fields and other agricultural and engineering structures by controlling their physical properties such as strength and weight of these kind of composite materials. For effective utilization of concrete or cementitious composites with abandoned shell aggregates in engineering structures and building components, it is necessary to study various engineering properties such as mechanical behavior, strength, weight control and permeability of these kinds of concrete or cement based composite materials.

The purpose of the work reported in this manuscript was to investigate the overall response of the mechanical properties of concrete or cementitious composites with different percentage of abandoned mactridae shell husk as coarse aggregate such as 0, 10%, 0, 30%, 40% and 50% in the ratio of mass. Various types of tests such as compression tests, permeability tests and durability tests on concrete containing different percentage of abandoned mactridae shell husk under severe environmental conditions for a period of 28, 88, 148, 208 and 268 days were demonstrated. A comparison of the test results and pertinent discussion regarding the investigated parameters are depicted in this manuscript.

2. Materials and Methods

2.1. Collection and Grading of Abandoned Shell Husk

The abandoned mactridae shell husks were collected from the sea coast near to Mie University, Tsu city, Mie Prefecture, Japan (**Figure 1**). After collection, the shell husks were graded by performing sieve analysis. The fineness modulus and the maximum size of the abandoned shell husks were 4.35 and 4.76 mm, respectively. The shell size distribution curve is shown in **Figure 2** and the physical properties are given in **Table 1**.

2.2. Specimen Preparation

The requisite amount of shells, sand and cement was dry-mixed in a pan, and then the requisite quantity of water was added gradually while the mix was continuously stirred (**Figure 3**).

Cylindrical specimens were made in the metal moulds with open tops. The moulds were made in such a way that the side walls and the base were detachable so that the mould could be easily separated from the specimens after its initial setting. The percentage of recycled shell

Figure 1. Abandoned Mactridaes hell husksin Tsu city, Mie Prefecture, Japan.

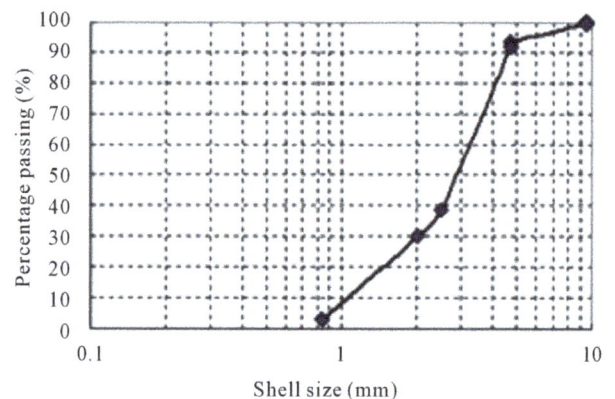

Figure 2. Shell size distribution curve.

Table 1. Physical properties of shell husk concrete.

Physical properties	Values obtained
Water absorption ratio	1.23%
Specific Gravity	2.723
Unit weight	1.57

Figure 3. Dry-mixing of shell husks with sand-cement.

aggregates were marked on the elements, and the date of casting were recorded. The contact surfaces of the mould to the mortar were greased before casting the specimens to ease the demolding process.

Ordinary Portland cement and river sand passing through a No. 8 (2.38 mm) sieve, which has a fineness modulus of 2.33, were used for casting. For all the specimens, the water to cement ratio and cement to sand ratio both were 0.5 by weight. The specimens were air-dried for 24 hours for initial setting (**Figure 4**) and then immersed in water for curing (**Figure 5**). The specimens were removed from water after 28 days and were air-dried for 2 days in room temperature of about 25°C and relative humidity of about 50%; then the tests were performed.

2.3. Testing under Compression

All the specimens were tested with a 1962 kN capacity hydraulic testing apparatus (**Figure 6**). The maximum capacity of the machine was adjustable to reduce scale and was set to 196.2 kN. The readings were taken initially at an interval of 9.81 kN and subsequently at 4.905 kN. The displacements were measured with a dial gauge having a least count of 0.01 mm. A careful attention was taken to eliminate the effect of end restraint of the test specimens. It was attempted to eliminate the initial slackness of the machine components and the contact points by exercising the test specimens under a small load before applying final load cycle to failure. The load application was continued until the deformation became excessive. Most of elements failed with cracking noise and some elements showed spalling off the mortar cover over the aggregates associated with a rapid drop in load.

2.4. Testing for Hydraulic Conductivity

Hydraulic conductivity testing apparatus used in this research is shown in **Figure 7**. The hydraulic conductivity tests were conducted by the constant head permeability test method according to JIS 1218. The specimens for the permeability tests were prepared in the steel mold of 10.0 cm in diameter and 20.0 cm in height. The specimen was cut to height 6.0 cm with permeable surfaces at top and bottom. Epoxy resin was used to block the clearance between the mold-wall and the specimen periphery so that water can pass within the specimen only. The mold with specimen was placed on a stand to ease in collecting the outlet water by a graduated cylinder. The water pressures gauge was set at the upper plate of the specimens to confirm the constant water pressure on the specimen. Before applying the pressure from the gas cylinder, the water from the water tank was poured to the pressure chamber by adjusting the water control valve. The water of pressurized chamber was then poured on the specimen. During pouring the water into the mold, the air inside the mold on the top of the specimen was released by the air

Figure 4. Casting of shell husk concrete.

Figure 5. Curing of shell husk concrete.

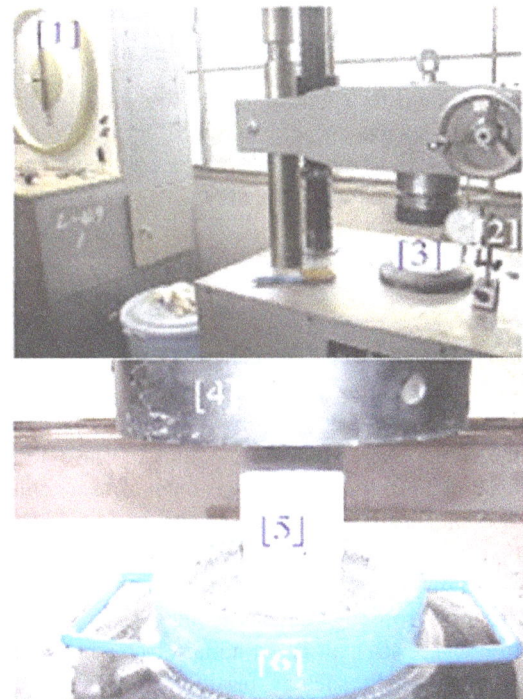

1) Load measuring dial gauge; 2) Displacement measuring dial gauge; 3) Lower part for application of load; 4) Upper part for reaction of applied load; 5) Specimen set up for testing and 6) Adjustment element for applied load.

Figure 6. Compressive test apparatus.

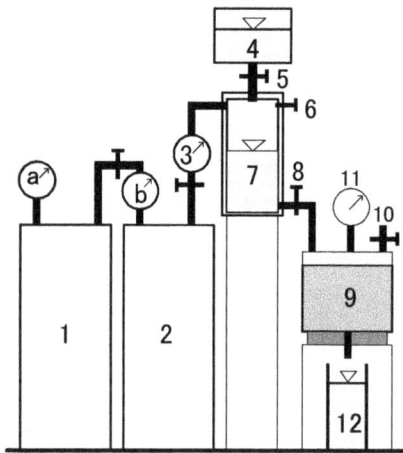

1. Gas cylinder (15 MPa); 2. Gas cylinder (3 MPa); 3. Control valve for gas; 4. Water supply; 5. Control valve for water supply; 6. Air releaser; 7. Pressurized water; 8. Control valve for pressurized water; 9. Specimen; 10. Air releaser; 11. Dial gauge; 12. Graduated cylinder.

Figure 7. Hydraulic conductivity test apparatus.

Table 2. Properties of shell husk concrete.

S	fc	γ	E	ft	w	k
0	30.00	2.31	12.78	3.10	4.0	0.0
10	26.33	2.22	10.23	2.95	13.4	1.1
20	20.67	2.20	8.14	2.73	19.7	9.7
30	18.13	2.00	6.82	2.69	26.2	88.9
40	15.33	1.99	5.93	2.57	34.2	150.2
50	10.97	1.94	5.07	2.28	39.1	403.4

Note: S = Shell content (%), γ = unit weight (t/m^3), fc = compressive strength (KPa), E = Young's modulus (MPa), ft = Tensile strength (KPa), w = Water absorption (g), k = Coefficient of permeability (m/s × 10^{-10}).

release valve. The front face of the specimen was continuously connected to the pressurized water chamber during the test. Therefore, the top surface of the water in the mold was subjected to a constant pressure of 3.0 MPa (gauge pressure). The amount of water flowing through the specimen was measured by the collected water in the graduated cylinder. Before recording the readings, percolation was allowed for some time to ensure a high degree of saturation and uniformity of the test results.

2.5. Tensile Strengths and Modulus of Elasticity

The test method for splitting tensile strength of cylindrical concrete specimens was followed according ASTM C496/C496M-11 standards. The size of specimen was length 20 cm and diameter 10 cm.

3. Results and Discussion

3.1. Results

The average values of 3 specimens of the cementitious composites containing 0, 10%, 20%, 30%, 40% and 50% recycled shell aggregate is given in **Table 2**. It is evident that the compressive strength, modulus of elasticity and tensile strength of the cementitious composites decreased with the increase in shell content. This table also shows that the unit weight of the composite decreased significantly with the increase in the shell content which can be taken as a controlling parameter along with strength. The structure where strength is not a main factor but it needs to be lighter weight, such as, partition wall; in that case, cementitious composites with higher percentage of shell content may be used. On the other hand, the structure where the weight is not a main factor but it needs to be strong and durable, such as, roads, embankment, pave-

ments and earth slope protection; in that case, cementitious composites with lower percentage of shell content may be used. Water absorption tests and hydraulic conductivity tests results show that the amount of water absorption and coefficient hydraulic conductivity (k) increased with the increase in the amount of shell content. The higher rate of water absorption and k was occurred due to the higher pore space in concrete. This indicates the effectiveness of the higher amount of shell contents in concrete.

The durability tests of concrete containing 50% shell treated in severe environment such as in saline water (sea-water) for a period 28, 88, 148, 208 and 268 days are given in **Figure 8**. Before placing the specimens in sea water, basic curing for a period of 28 days in fresh water at the laboratory was done for all the specimens. This figure shows the average compressive strengths of 3 specimens in each group. Notice a significant increase in strengths of concrete due to the presence of shell husk aggregate indicating the effectiveness of shell husks in concrete treated in saline water (sea-water).

3.2. Interrelationships of Mechanical Properties

By analyzing experimental data, interrelationships among the physical properties of concrete containing different amount of shell husk such as relationships of the compressive strengths and the modulus of elasticity, relationships of the tensile and the compressive strengths are given as follows.

3.3. Relationships of the Young's Modulus and the Compressive Strength

According to the test method of ASTM C469, the laboratory tests were carried out and the Young's modulus was calculated from the stress strain relationships by avoiding the nonlinearity at the initial and final loading conditions using Equation (4).

$$E = \frac{\sigma_2 - \sigma_1}{\varepsilon_2 - 0.00005} \tag{1}$$

where E is the modulus of elasticity, σ_1 is the stress corresponding to strain of 0.00005, σ_2 is the stress corre-

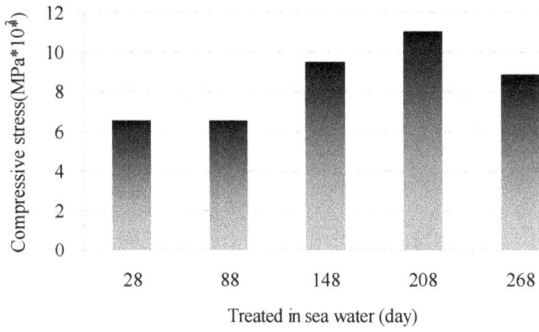

Figure 8. Durability tests in saline water.

sponding to 40% of ultimate strength and ε_2 is the strain produced by stress σ_2. The following relationships between the compressive strengths and the Young's modulus are obtained based on the experimental data:

$$E = \alpha \left(f_c' \right)^{1/2} \tag{2}$$

where E is Young's modulus of composite in MPa, f_c' is the compressive strength of composite in MPa, α is the constant which has values of 2333, 1993, 1790, 1601, 1566 and 1530 for concrete of control, 10%, 20%, 30%, 40% and 50% shell husk, respectively. It is noted here that the empirical relationship between the compressive strength and the Young's modulus of recycled aggregate concrete by the ACI code is given as follows:

$$E = 33\gamma^{3/2} \left(f_c' \right)^{1/2} \tag{3}$$

where γ is the unit weight of concrete in pound per cubic feet. This Equation is the well-known form for normal weight concrete with γ value of 145 pound per cubic feet, which is written as follows:.

$$E = 57000 \left(f_c' \right)^{1/2} \tag{4}$$

The Equation 4 by the ACI code can be rewritten in SI unit as follows:

$$E = 4734 \left(f_c' \right)^{1/2} \tag{5}$$

where, Young's modulus (E) and compressive strength (f_c') are in MPa. The α values of Equation 2 obtained in this research are smaller than the coefficient of Equation 5 given by ACI indicates the proposed Equation provides conservative estimation of the Young's modulus of concrete containing shell husks. This type of behavior was observed in the technical literature in case of concrete containing different recycled aggregate [11-13].

3.4. Relationships of the Splitting Tensile Strength and the Compressive Strength

By performing the analyses of the experimental data, the relationship between the splitting tensile strength and the compressive strengths can be written as follows:

$$f_t = \beta \left(f_c' \right)^{1/2} \tag{6}$$

where, f_t is the splitting tensile strength in MPa, f_c' is compressive strength in MPa and β is an empirical constant which takes the values of 0.56, 0.57, 0.60, 0.63, 0.67 and 0.69 for concrete of control, 10%, 20%, 30%, 40% and 50% shell husk, respectively. It should be pointed out here that the ACI code used for estimating the splitting tensile strength is given follows:

$$f_t = 0.56 \left(f_c' \right)^{1/2} \tag{7}$$

where, f_t is the splitting tensile strength in MPa and f_c' is compressive strength in MPa. It is evident that the results agree well with the ACI equation of recycled aggregates concrete of plain mortar.

3.5. Discussion

The scanning electron microscope (SEM) analysis was done on the shell husks concrete and SEM image is given in **Figure 9**. As can be seen, the SEM image has cavity or pore space inside the shell husk concrete. This pore space facilitate the water absorption capacity of shell husk concrete which further reinforced the results given in **Table 2**.

The initial and final nonlinearity was avoided from the stress strain relationships. Although there was a limitation of test setup and the displacement measuring gage was set at machine component, however, the axial strain obtained during testing was similar to that of change in height or length of the specimen and the results obtained are promising for abandoned shell husk concrete. The tensile splitting tests were carried out on the cylindrical specimens. The test method for splitting tensile strength of cylindrical concrete specimens was followed according ASTM C496/C496M-11 standards. The size of specimen was length 20 cm and diameter 10 cm. The stress displacement curve is shown in **Figure 10**.

After eliminating the initial and final nonlinearity, the linear portion shown **Figure 10** is used to calculate to modulus of elasticity given in **Table 2**. The stress-strain curves for 10% shell husk concrete are reported. The stress-strain curves for other cases are similar trend and therefore, these are not repeated here. It is noted here that the slump tests were performed for the workability of shell husk mortar mix. For all the mixes, the value of slump was found as 70 - 95 mm without any admixture. It is interesting to note that the high water absorption of shell husk concrete reduces the rain water retention on the surface of the structure made of shell husk concrete. Therefore it reduces the risk of flood water during rainfall. This is beneficial for earth slope protection work to reduce pore water pressure inside the slope and thereby increases the stability of slope.

4. Conclusion

The possible utilization of abandoned shell husks in con-

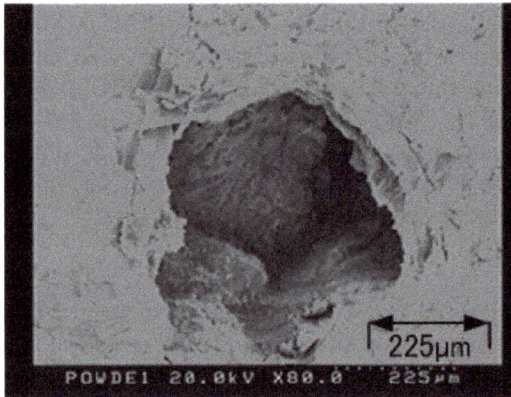

Figure 9. Scanning Electronic Microscopic (SEM) image of shell husk concrete.

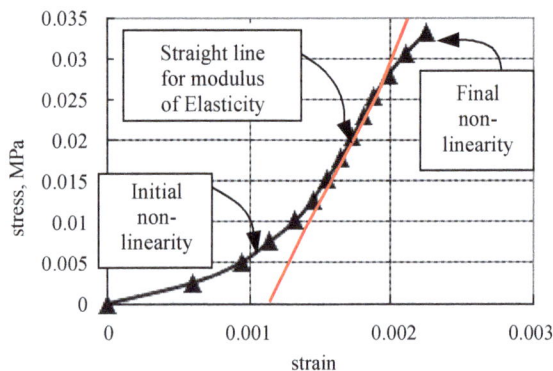

Figure 10. Stress-stress curve and linearity for modulus of elasticity.

crete construction is studied in this paper. Engineering properties such as compressive strength, Young's modulus, tensile strength, unit weight, water absorption capacity and coefficient of hydraulic conductivity results agree well with the ACI results. In designing light weight concrete structures such as partition wall, slope surface protection, ridge between paddy fields, roads and embankments slopes and sea-shore protection structures where strength is not a great factor but the water absorption properties are advantageous, concrete with shell aggregates may be used.

5. Acknowledgements

The present study is partly supported by the Research Grant No. 22580271 with funds from Grants-in-Aid for Scientific Research, Japan. The author gratefully acknowledges these supports. Any opinions, findings, and conclusions expressed in this paper are those of the authors and do not necessarily reflect the views of the sponsor.

REFERENCES

[1] I. Wang, S. Takamura, M. Nakamura and Y. Tsukinaga, "A Study on the Ratio and Composition of the Porous Concrete Containing Shells," *Transactions of Cement and Concrete*, Vol. 57, 2003, pp. 572-577 (In Japanese).

[2] M. Delwar, M. Fahmy and R. Taha, "Use of Reclaimed Asphalt Pavements as an Aggregate in Portland Cement Concrete," *ACI Materials Journal*, Vol. 94, No. 3, 1997, pp. 251-256.

[3] Y. Kansai, "Demolition and Reuse of Concrete Masonry," *Proceedings RILEM Symposium*, Tokyo, Chapman and Hall, 7-11 November 1988, p. 774.

[4] T. C. Hansen, "Recycled Aggregates and Recycled-Aggregates Concrete, State-of-the-Art Report Developments 1945-1985," *RILEM Technical Committee 37-DRC, Material Structures*, Vol. 19, No. 11, 1986, pp. 20-46.

[5] K. A. Rim, A. Ledhem, O. Douzane, R. M. Dheilly and M. Queneudec, "Influence of the Proportion of Wood on the Thermal and Mechanical Performances of Clay-Cement-Wood Composites," *Cement and Concrete Composites*, Vol. 21, No. 4, 1999, pp. 269-276.

[6] A. Ajdukiewicz and A. Kliszczewicz, "Influence of Recycled Aggregates on the Mechanical Properties of HS/HPC," *Cement and Concrete Composites*, Vol. 24, No. 2, 2002, pp. 269-279.

[7] N. Banthia and C. Chan, "Use of Recycled Aggregate in plain and Fiber Reinforced Shot Crete," *Concrete International: Design and Construction, American Concrete Institute*, Vol. 22, No. 6, 2000, pp. 41-46.

[8] K. K. Sagoe-crentsil, T. Brown and A. H. Taylor, "Performance of Concrete Made with Commercially Produced Coarse Recycled Concrete Aggregate," *Cement and Concrete Research*, Vol. 31, No. 5, 2001, pp. 707-712.

[9] M. Z. Hossain and S. Inoue, "Finite Element Analysis of Thin Panels Reinforced with a Square Mesh," *Journal of Ferrocement*, Vol. 32, No. 2, 2002, pp. 109-125.

[10] D. C. Stahl, G. Skoraczewski and B. Stempski, "Light Weight Concrete Masonry with Recycled Wood Aggregate," *Journal of Materials in Civil Engineering*, Vol. 14, No. 2, 2002, pp. 116-121.

[11] T. Yamauchi, H. Sahara and K. Kudo, "On the Use of Hotate Shells as Concrete Aggregates," *Proceedings of the 60th Technical Conference of Civil Engineering*, Tokyo, 7-9 September 2005, pp. 435-437 (in Japanese).

[12] S. T. Frondistou-Yannas, "Waste Concrete as Aggregate for New Concrete," *ACI Journal*, Vol. 74, No. 8, 1977, pp. 373-376.

[13] T. C. Hansen and E. Boegh, "Elasticity and Drying Shrinkage of Recycled Aggregate Concrete," *ACI Journal*, Vol. 82, No. 5, 1985, pp. 648-652.

10

Understanding the Occurrence of Two Total Floats in One Activity and Schedule Crashing Approaches for That Situation

Boong Yeol Ryoo[*], **Mike T. Duff**

Department of Construction Science, Texas A&M University, College Station, USA.

ABSTRACT

Critical Path Method (CPM) Scheduling has proven to be an effective project management tool. However, teaching the topic has proven it difficult to include all elements of CPM yet keep it simple enough for students to understand. In an effort to simplify the teaching of critical path method scheduling, the issue of two total floats in an activity does not get the attention necessary to address its occurrence. The objective of this paper is to present a mathematical method to show multiple total floats are possible for an activity. Also presented are suggestions for schedule crashing when multiple total floats are found. Two totals floats can be found if constraints (Lag or Lead) or non-Finish-to-Start (FS) relationships, or both are used in a network diagram. Situations are possible where an activity may have a start total float (STF) of zero but have a finish total float (FTF) greater than zero, or vice versa. Because the critical path generally follows the zero total float, these situations, where either the STF or the FTF is critical while the other is not, determine how the critical path activity must be controlled and crashed. This paper will present approaches of how to crash the schedule when a portion of the activity, either start or finish, is critical. Also presented will be methods to teach the subject matter with or without the use of scheduling software. Critical Path Method was revisited to see what the minimal conditions are needed to have activities with two total float. Generalized crashing methods were studied to see if the methods can be used when two total floats exist.

Keywords: Critical Path Method; Multiple Total Floats; Schedule Crashing; Constraints; Lag/Lead

1. Introduction

1.1. Background

The construction schedule is a communication tool at the construction site [1]. Due to the fact that specialty contractors execute contractually responsible tasks for a project and they may not be able to work together directly, a well-developed construction schedule can be used to monitor construction operations, project resources, potential risks, and desired quality of the project. The schedules are made up of a list of deliverables to be completed, construction activities required for the deliverables, and administrative activities to support the construction activities [1]. The administration activities are the project managers' or superintendents' activities to the schedule to oversee the management responsibilities for general contractor's or specialty contractors' construction operations.

The network diagramming method is commonly used in building construction. Typically four activity relationships are used to represent a sequence or constraint between two activities: Finish-to-Start, Finish-to-Finish, Start-to-Start, and Start-to-Finish [2]. The critical path method is widely used for project planning in the commercial construction industry to find the longest path of activities to complete a project. First, four different dates such as early start date, early finish date, late start date, and late finish date for each activity are calculated using the Forward Pass and Backward Pass methods. Then, at least two floats, total float (TF) and free float (FF) for each activity, are calculated in order to identify if the activity can be delayed without delaying the whole project or immediate following activities. Finding zero TF activities is the first key to identify critical activities because FF of the critical activities is also zero. Commonly TF for an activity is calculated by determining the difference between the early dates or late dates of the activity assuming that the start TF and finish TF of an activity are the same. This calculation has been used even though multiple relationships are used between two activities. However, if multiple relationships or lead/lag time is used, the start TF and finish TF of an activity may not be the same. Related to construction scheduling, Feigenbaum [3] and Mubarak [4] briefly stated a possibility of multiple TF for an activity while others did not explain the possibility [3-17]. Thus, the start and finish dates can be calculated but the floats cannot be calculated accurately under the above conditions.

In scheduling real world construction projects multiple relationships are often used between activities. Lag or lead times are often used to show necessary time gaps in these relationships. In addition, constraints are also frequently added to impose limitations on activities. These could add another level of complexity to the schedule because the start TF and the finish TF of an activity can be different. When this happens, only the beginning or ending point of the activity becomes critical. This paper is to investigate conditions which cause two total floats for an activity and the affect this has on crashing the schedule.

1.2. Observation

When multiple relationships or a lag/lead time is used in a CPM diagram a situation can arise where there are two different total floats for an activity, start total float (STF) and finish total float (FTF). STF is the amount of delay at the beginning of the activity whereas FTF is the amount of delay at the end of the activity without delaying the whole project. When this happens, either the start or finish dates of the activity, whichever is critical, must be followed because the critical path follows zero total floats. It affects how a project can be managed based on where the critical path follows.

In order to find the maximum time reduction with the minimum cost increase, the activity with the cheapest crashing cost on the critical path of the project is always selected [1]. When STF and FTF are different, shortening activity durations may not be enough to reduce its duration because relationships/constraints dictate the start or finish date of the activity. Thus the current critical path crashing methods must be reexamined to determine if this is taken into consideration and acted upon.

2. Determining Critical Paths

2.1. Current Critical Path Calculation Methods

The critical path method is widely used to find the critical paths and critical activities in network schedules.

Through the performing of the forward pass calculation the early start date (ESD) and early finish date (EFD) for an activity is calculated. The late start dates (LSD) and late finish date (LFD) of each of the activities is calculated by performing the backward pass calculation. Once the ESD, EFD, LSD, and LFD have been calculated then total float and free float can be determined as in Equation (1). The calculation of total float for the activity is crucial because the critical path follows the activities of zero total float [6].

$$TF = LSD - ESD \text{ or } LFD - EFD \qquad (1)$$

Based on the assumption that there is only one total float for the activity, scheduling programs adopt Equation (1) and only one total float for each activity is calculated. This leads to another assumption that the total float at the beginning and end of the activity is the same. It is a safe solution but not enough to effectively reduce the duration of the project because real amount of slacks are not known.

With these assumptions there is the risk of missing the two total floats for the activity. Along these same lines Mubarak [4] and Marchman and Sulbaran [13] introduced an additional equation to calculate total float for an activity as used in Equation (2). This equation, in addition to Equation (1), still only calculates one total float for the activity, but does begin to compare the difference between the start and finish of an activity.

$$TF = LFD - Duration - ESD \qquad (2)$$

Feigenbaum [3] puts forth two equations where two total floats are calculated; start total float and finish total float for an activity as shown in Equation (3) and Equation (4).

$$STF = LSD - ESD \qquad (3)$$
$$FTF = LFD - EFD \qquad (4)$$

Comparing these two equations serves as a way to determine if there is an instance of two different total floats for an activity.

In the event that Equation (3) and Equation (4) are equal then this means that STF and FTF are equal. Without a lad/lead time, durations are the only factor affecting the project duration.

Free float is determined by Equation (5) and it shows how much the activity can be delayed without delaying the following activities. This cannot be used if a relationship other than finish-to-start is used. In this case, the early start date for the activity may be irrelevant because these relationships represent direct relationships between two activities not effect the early start.

$$FF = minimum\left(ESD_{successors}\right) - EFD \qquad (5)$$

2.2. Existence of Two Total Floats

Between any two activities, there can be multiple paths

from the first activity to the last activity. The duration of any path between two activities is the sum of the durations and lag/lead times of the activities in the path. Under normal circumstances the largest early finish date is the early start date for the immediate successor activity. However, multiple relationships, such as the addition of lag and lead times, between two activities may cause this to not be the case. The Early Start for the activity I can be calculated using Equation (6).

$$Duration \; of \; Path_i = \sum_{j=0}^{k} Duration_j + Lag_j \qquad (6)$$

where, i = any path between any two activities;

j = Activity number;

k = the total number of activities of each path.

As stated before, with no lags the early start date of an activity is the largest early finish date of the paths before the activity. However, when lags are used, ESD, EFD, LSD, and LFD can be calculated as shown using the following equations and demonstrated in **Figure 1**.

$$ESD = EFD_i + Lag_1 \qquad (7)$$

$$EFD_j = EFD_i + Lag_1 + Duration_j \qquad (8)$$

$$EFD_j = EFD_i + Lag_2 \qquad (9)$$

$$LSD_j = LSD_k - Lag_4 \qquad (10)$$

$$LFD_j = LSD_k - Lag_2 \qquad (11)$$

$$LFD_j = LSD_k - Lag_4 + Duration_j \qquad (12)$$

Equations (13) and (14) show the distance of two connected activities. In order to maintain the same distance between two activities, Equations (13) and (14) must be true. If Equations (3) and (4) yields two different total floats, the start and late dates can be calculated so that the calculation of STF and FTF become possible.

$$Lag_1 + Duration_j = Lag_2 \qquad (13)$$

$$Lag_4 - Duration_j = Lag_3 \qquad (14)$$

In order to maintain the same duration for Activity j, the duration can be calculated from two linked lags as in Equation (15).

$$Duration_j = Lag_2 - Lag_1 = Lag_4 - Lag_3 \qquad (15)$$

In **Figure 1**, STF and FTF are different if $Lag_2 - Lag_1 \neq Lag_4 - Lag_3$. On the other hand if the following equations, Equations (16) and (17), are true, then the possibility of different STF and FTF for the activity exists.

$$Duration_j \neq Lag_2 - Lag_1 \qquad (16)$$

$$Duration_j \neq Lag_4 - Lag_3 \qquad (17)$$

Thus,

$$LSD - ESD \neq LFD - EFD \qquad (18)$$

This leads to the conclusion that slacks at the beginning and end of the activity can be different as in Equation (18). Thus, as a lag/lead time is used between two activities, there is a possibility that either the start or finish date can only be critical. It is also true if multiple relationships are used to link two activities as in **Figure 1**.

3. Crashing the Schedule with Two Total Floats

3.1. Working with a Two Total Float Example

Figure 2 is a CPM diagram with multiple relationships and lags. While it is a nine activity example it serves to demonstrate what can occur when there are multiple relationship types and lags.

In **Figure 2** the following are the critical activities:

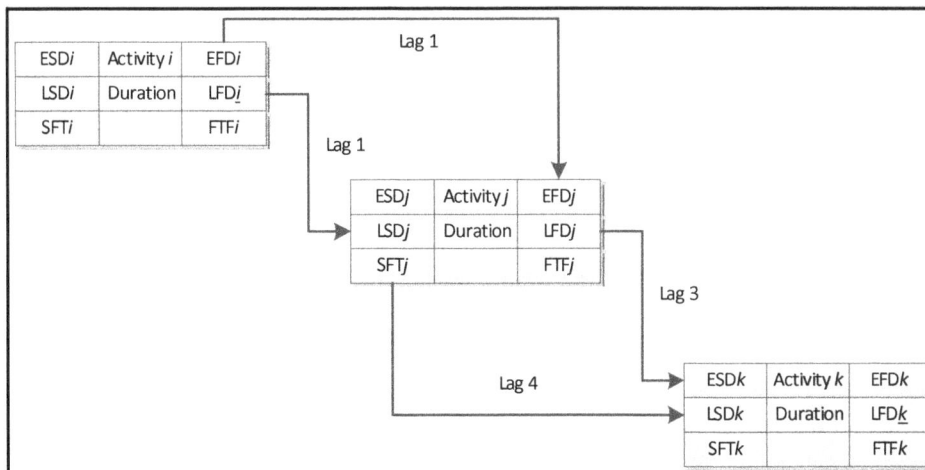

Figure 1. Diagram of lags and relationships.

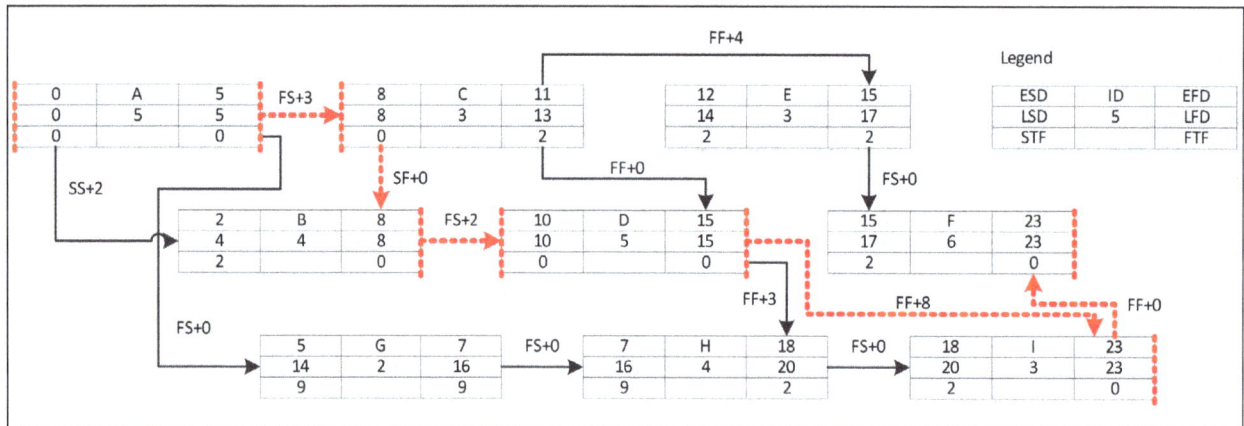

Figure 2. Two total float example with multiple relationships and lags.

- Activity A (enitre).
- The Start of Activity C.
- The Finish of Activity B.
- Activity D (entire).
- The Finish of Activity I.
- The Finish of Activity F.

The project duration is 23 days and is as follows:

- All of Activity A is critical; meaning both the start date and finish date must be kept to avoid a delay in the entire project.
- The finish date of Activity B must complete on time, but the start has 2 days of float. Thus, available maximum total float of Activity B is 2.
- The start of Activity C has zero total float and the finish has 2 day of float.
- Both the start and finish dates for Activity D are critical with zero days of float for both.
- Activity E while it is not critical has a different STF of 2 days and a FTF of 2 days. Thus, available maximum total float of Activity E is 2.
- While Activity F must finish on day 23 it has 2 days of start float. Thus, available maximum total float of Activity F is 2.
- Activity G is a non-critical activity and it has identical values for both STF and FTF. It also is the only activity that has a finish-to-start relationship with both its predecessor and successor.
- Activity H's STF is 9 days and its FTF is 2. Thus, available maximum total float of Activity B is 7.
- The finish of Activity I has zero days of float while the start date has 2 days of float. Thus, available maximum total float of Activity I is 2.
- Activities B, C, I and F are governed by their non-FS relationships.

With a hand drawn schedule, similar to the Activity on Node in **Figure 2**, it is possible to find these incidences of a different STF and FTF. The red hidden lines represent critical paths of the schedule. Vertical red hidden lines indicate critical dates, either start or finish dates. A few activities are reviewed as follow:

STF of Activity B is 2 and its FTF is 0. If the activity starts on the early date, the activity can have as much as 2 days of total float at the beginning because the activity must be completed by Day **8**.

STF of Activity E is 2 and its FTF is 2. Activity E may not have 2 days of total float unless it begins on the early start date. If it is delayed more than 2 days, which is the amount of its FTF, then early dates are no longer feasible.

In the same manner, the TF of Activity H is only 7 even through STF of Activity H is 9 because there is a two day lag at the end of Activity H. It is required to begin Activity H on its early start date to maximize its slack of 9. Otherwise the TF of Activity H will be reduced to 2. Utilization of 9 day delay on Activity H at the beginning forces to use the late schedule for the rest of the project.

3.2. Comparison of Scheduling Software

Two scheduling softwares, Microsoft Project (MSP) and Primavera P6 Project Manager (P6), were used to compare the results of the same schedule. As shown in **Tables 1** and **2**, Microsoft Project® and Primavera 6® show different results than the hand calculation in **Figure 2**.

The main issue is the algorithm written by the programmers to deal with the critical path calculations of the logic. For instance in MSP while there is a way to show both STF and FTF, MSP does not show the differences in activities B, C, E, F, H, and I found in the hand calculated diagram. With the scheduling programs, early start and early finish dates are less flexible than hand calculation if there is a lag or non-Finish-Start relationship such as Activities B, E, H, and F. The programming of these portions of the software has not been set up to accept all the implications of these different relationships and lags. This eliminates some potential analysis from the beginning, because MSP only allows one relationship between

Table 1. Schedule data and activity dates from Microsoft project sort by early start date.

ID	Task Name	Duration (days)	Predecessors	Early Start Date	Early Finish Date	Late Start Date	Late Finish Date	Total Float	Free Float
1	Total	23	-	1	23	1	23	0	0
2	A	5	-	1	5	1	5	0	0
3	B	4	2SS + 2.4SF	5	9	5	9	0	0
8	G	2	2	6	7	15	16	9	7
4	C	3	2FS + 3	9	11	9	11	0	0
5	D	5	3FS + 2.4FF	11	15	11	15	0	0
6	E	3	4FF + 4	13	15	15	17	2	2
9	H	4	8.5FF + 3	15	18	17	20	2	2
7	F	6	6.10FF	18	23	18	23	0	0
10	I	3	5FF + 8.9	21	23	21	23	0	0

Table 2. Schedule data and activity dates from primavera 6 sorts by early start date.

Task Name	Duration (days)	Predecessors	Early Start Date	Early Finish Date	Late Start Date	Late Finish Date	Total Float	Free Float
Total	23	-	1	23	1	23	0	0
A	5	-	1	5	1	5	0	0
B	4	A, C	5	9	5	9	0	0
G	2	A	6	7	15	16	9	9
C	3	A	9	11	8	11	0	0
D	5	B, C	11	15	11	15	0	0
E	3	C	13	15	15	17	2	2
H	4	D, G	15	18	17	20	2	2
F	6	E, I	18	23	18	23	0	0
I	3	H, D	21	23	21	23	0	0

two activities.

Because of this hand calculation methods and scheduling programs must be taught together so students can experience the difference of the results from hand calculation and schedule programs. Also in the industry it is a way to double check the logic of a schedule. **Figure 2** shows that the focus on the activities affected by non FS or activities with Lags is all that is necessary to determine if there are differences in STF and FTF. In other words, it is not necessary to hand draw the logic of the entire schedule, just those activities affected by lags and non-finish-to-start relationships.

As shown in **Tables 1** and **2**, totals for Activities E, G, and H from both programs are the same. Activity B has a 2day SFT but it was not included in the calculation because the finish date of Activity B is critical. Reversely Activity C's FTF is 2. Activity C is forced to obey the early start date because it was critical. However Activity F's finish date is critical so it is forced to follow the late dates.

Activity G demonstrates the differences in algorithms between MSP and P6. The relationship between G and H

is a finish-to-start with no lag or lead. The total float for G was calculated correctly, however, for free float it shows their way of trying to deal with the difference between the STF and FTF of Activity H. The free float given to Activity G does not belong to it. However, because both P6 and MSP do not know what to do with the situation in Activity H they default to what they can do. MSP does take into consideration the FTF of two days for Activity H, by taking it away from the Free Float listed for Activity G.

3.3. Crashing the Schedule

To achieve the least expensive duration reduction, or crashing, of the schedule is to find the cheapest and shortest option the duration of the project can be reduced. This is a fairly common occurrence on projects. The following steps are generally accepted to reduce the duration of the project to meet the revised end date. In order to find the minimum cost schedule, trade-off analysis is necessary to find the lowest cost alternatives in each step. Generally the lowest crashing cost activities are selected and crashed. Brunnhoeffer and Celik [2] presented a

general algorithm for crashing the schedule.
- Identify the critical path.
- Select the cheapest crashing cost activity.
- Crash the selected activity 1) to the next longest path; 2) until the critical path changes; or 3) until crashing is infeasible. The first approach is preferred because it does not crash the schedule more than necessary.

This approach can be used if the early total float and finish total float for an activity are the same. However, in the schedule in **Figure 2** if Activity C were to be shortened because it is listed as critical according to both P6 and MSP, it would not have an effect on the end of the total project because only the start of Activity C is critical. Relationships that are not FS or have Lags do not lend themselves to traditional crashing methods.

Table 3 lists the activities found in **Figure 2** with their data needed for crashing. While the activities listed in **Tables 1** and **2** are on the critical path only two activities are entirely critical, Activities A and D. The start date of Activity C is critical whereas the finish dates of Activities B, F, and I are critical. Reducing the duration of B does not affect the total project duration because the relationship from Activity C to Activity B has a greater effect than that of Activity A to Activity B. Activity F, which is on the critical path and at $300 per day is the cheapest activity to crash, can be reduced by 2 days. The total project duration will not be reduced because of its constraint from Activity I. Thus it is recommended to see if the lag from Activity I can be revised. The relationship between Activity I and Activity F is Finish-to-Finish (FF) with zero lag. It cannot be crashed. However it will have more total float if it is crashed. The next choice is Activity B, $500 per day; it can be reduced by 1 day. However, because Activity B has a constraint enforced by Activity C to the finish date of Activity B, the project duration will not be affected. Activity C's start date is critical and it is the next alternative, $700 per day. Reducing the duration of Activity C may or may not be beneficial to the project. Reducing its duration can affect the finish date of Activity D which happens to be critical. Otherwise it may not affect the project duration. However, since the entire activity of Activity A and Activity D have the same priority, $1000 per day, reducing their duration can affect the duration of the project.

Another example of the effects of non FS relationships is seen in **Figure 3**, which is a portion of the logic diagram for placing a footing. The inspection needs to be completed on Day 13, however, it can start as early as Day 10. In other words, according to the logic, there needs to be an inspection as the rebar and formwork is being placed, there also needs to be an inspection done when the Footing concrete is placed. This problem would best be solved by splitting the inspection activity into two separate activities to take care of both the needs spelled

out by the logic. This example does serve to demonstrate some of the issues that can occur with the use of lags and non-finish-to-start relationships to take care of some logic needs of a project.

As shown in **Figure 2**, some activities in the critical paths are critical based on either their start date or finish date. Conventional crashing guidelines cannot be applied to reduce the project duration because it assumes the whole activity is critical [2,12]. Based on the guideline, reduction in activity duration on the critical path causes automatic reduction in schedule time. However, if either only the start date or finish date is critical, crashing a critical activity does not cause an automatic duration reduction in the schedule as.

When this is the case, especially when STF of an activity is critical, a revision of relationships to the activity is suggested prior to crashing the activity. The following additional suggestions should be considered:
- Identify the critical path.
- Reduce lag times affecting the critical path.
- Revise relationships from finish-to-start, start-to-finish, or finish-to-finish to start-to-start relationships.
- Select the cheapest crashing cost activities and 1) see if any relationship can be crashed if the finish date of the activity is only critical; 2) do not attempt crash it if it is not; and 3) do not crash the activity if only the start date is critical;
- Crash the least expensive activity if both start date and finish date are critical.

4. Suggestions

When constraints are used in a project, this creates the need to use non FS relationships and lags and leads. This leads to two observations. The first observation is if lags and leads are used there is a possibility of the STF being different than the FTF. Secondly, multiple relationships between activities can also create these differences. Also, other constraints not placed by management or the owner can also lead to the difference between STF and FTF.

When this is the case, calculations of STF and FTF are recommended. It is only necessary to do these calculations on those activities affected by the lag/lead or non FS relationship. In such cases, it is recommended these be done by hand and not be dependent on the scheduling software to tell if the start date, finish date, or entire activity is critical.

Secondly, before choosing a method to crash the schedule with an activity with two total floats it must be determined if the start date is critical or the finish date. If the start date is critical the predecessors need to be looked at to determine if the relationship or the activity is critical. In the event of two total floats the relationship should be crashed. If it is a lag relationship the lag needs

Table 3. Schedule data for crashing.

Activity	Normal Duration (days)	Crash Duration (days)	Normal Cost ($)	Crash Duration ($)	Maximum Reduction (days)	Cost to crash per period ($/day)	Critical Dates
A	5	3	5000	3000	1	1000	Start Date, Finish Date
B	4	3	2000	1500	1	500	Finish Date
C	3	2	2100	1400	1	700	Start Date
D	5	3	5000	3000	2	1000	Start Date, Finish Date
E	3	3	1800	1800	0	0	-
F	6	4	1800	1200	2	300	Finish Date
G	2	2	4000	4000	0	0	-
H	4	3	1600	1200	1	400	-
I	3	3	1500	1500	0	0	Finish Date

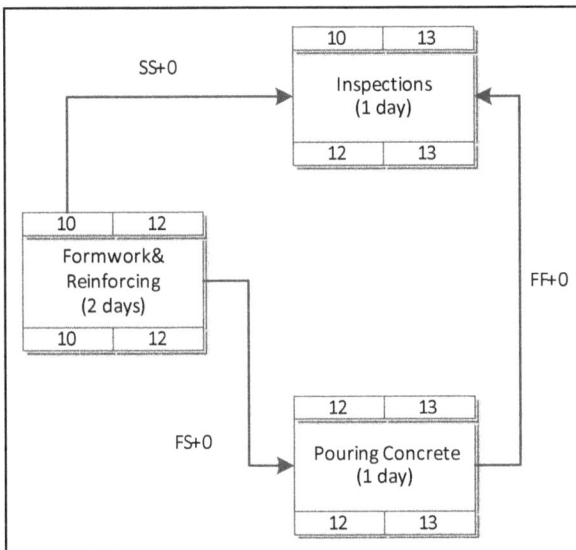

Figure 3. Concrete work logic (example used with permission).

to be crashed before the duration is crashed. It may not add any additional costs to the project unlike a duration crash.

Third, the generalized crashing method cannot be used to crash the schedule if non finish-to-start relationships and lag/lead times are used. The following is suggested: if the finish date is critical, reduce the amount of lag. The relationship and lag may control the activity itself. If the start date is critical, reducing its duration may or may not affect the project duration. The critical path should be recalculated in order to find any impact.

Lastly, understanding hand calculation methods and how scheduling programs work is important in order to create realistic schedules. People involved in this area need to understand how construction logic works, as well as understand how scheduling software uses that logic to create the schedule on the screen. The decisions that are made based on a computer generated schedule have se-

rious cost and time implications to a project, care should be taken to ensure the decisions are being made with correct and relevant information.

REFERENCES

[1] AGC, "Construction Planning & Scheduling," AGC of America, Arlington, 1997.

[2] G. C. Brunnhoeffer III and B. G. Celik, "Crashing he Schedule—An Algorithmic Approach with Caveats and Comments," 2010.

[3] J. Buttelwerth, "Computer Integrated Construction Project Scheduling," Prentice Hall, Upper Saddle River, 2005.

[4] D. G. Carmichael, "Project Planning, and Control," Taylor & Francis, London, 2006.

[5] F. E.Gould, "Managing the Construction Process," Prentice Hall, Upper Saddle River, 2005

[6] L. Feigenbaum, "Construction Scheduling with Primavera Project Planner," 2nd Edition, Prentice Hall, Upper Saddle River, 2002.

[7] S. Mubarak, "Construction Project Scheduling and Control," John Wiley and Sons, Hoboken, 2010.

[8] T. E. Glavinich, "Construction Planning & Scheduling," AGC of America, Arlington, 1997.

[9] T. E. Glavinich, "Construction Planning & Scheduling Manual," AGC of America, 2004.

[10] T. Hegazy, "Computer-Based Construction Project Management," Prentice Hall, Upper Saddle River, 2002.

[11] J. W. Hinze, "Construction Planning and Scheduling," 4th Edition, Prentice Hall, Upper Saddle River, 2011.

[12] H. Kerzner, "Project Management: A Systems Approach to Planning, Scheduling, and Controlling", 10th Edition, John Wiley & Sons, Inc., Hoboken, 2009.

[13] D. A. Marchman and T. A. Sulbaran, "Scheduling with Suretrak," 2nd Edition, Delmar Cengage Learning, Independence, 2006.

[14] J. S Newitt, "Construction Scheduling: Principles and

Practices," 2nd Edition, Prentice Hall, Upper Saddle River, 2011.

[15] J. J. O'Brien and F. L. Plotnick, "CPM in Construction Management," McGraw Hill, Boston, 2009.

[16] T. E. Uher, "Programming and Scheduling Techniques", University of New South Wales Press Ltd., Sydney, 2003.

[17] S. C. Weber, "Scheduling Projects Principles and Practices", Prentice Hall, Upper Saddle River, 2005.

11

Interpretable Housing for Freedom of the Body: The Next Generation of Flexible Homes

Kyung Wook Seo[1], Chang Sung Kim[2]*

[1]Department of Architecture, Kyonggi University, Suwon, South Korea; [2]Departmentof Architectural Engineering, Hyupsung University, Hwaseong, South Korea.

ABSTRACT

If we have gone through the first generation of housing design that pursued functional optimization, ergonomics, and circulation efficiency during the last century, now we are living in the second generation where more advanced goals, such as universal design, ubiquitous design, sustainable design, and environment-friendly design, are emphasized. Although this second generation of design focuses upon the wellness of humans in accordance with environment, it still has the attitude that a more precisely designed home can guarantee a better life. What lacks in this approach is the freedom of the body; it needs to make its own choice as to how to use a space. Thus, it is suggested in this paper that what is important in designing a home is to provide alternatives in daily lives so as to make a full exploration of a given space. These alternatives can be made by offering residents an interpretable space where they can figure out space usages and routs in a constantly changing context. Two spatial devices are discussed in depths as a way to realize this interpretable house: room-to-room enfilade and ring spatial structure. By investigating some existing house plans, it is illustrated how they can guarantee the freedom of the body, and thus alternatives for the flexible domestic life.

Keywords: Interpretable House; Bodily Freedom; Flexibility; Polyvalency

1. Introduction: Third Generation of Healthy Home Movement

The modernist movement in the twentieth century has left a functionalistic legacy that emphasizes the optimal programming of architectural space. Following this trend, the house planning in particular has moved towards a scientific realm where precise analysis and anticipation of economical human movement are sought. Technology has accelerated this movement further, making more efficient building environment that could cut away useless junks of space and combine functions by offering electrical and mechanical devices and installations. In the global perspective, this modern movement of functional optimization, ergonomics, and circulation efficiency has made domestic space more tightly integrated in most parts of the developed world. The UK, for example, has experienced the typical integration process of functional spaces in the early twentieth century; thus, the old spatial division of parlor/living room and dining room/kitchen, which separated formal and public activities from informal and private ones, has given way to a combined func-

tional space of living-parlor and dining-kitchen equipped with automated home appliances. The main concern in this modernized home was "running the house rather than social proprieties" and the whole house potentially became a clean display zone [1]. The concept of flexibility and freedom in domestic living was constantly discussed in both practice and research; but it was oriented towards the "rationalistic, scientific ethos of the age" [2].

If we label the above trend of house planning—mainly focusing on efficiency—as the past generation of healthy home design, then what could be the defining characteristics of the current house design movement? The followings are the current issues in the housing design sector:

Universal Design.
Ubiquitous Design.
Sustainable Design.
Environment-friendly Design.

Some issues seem to be a bit older than the others, but still all of these have become popular topics in the housing industry within the last few decades. They differ in their meanings and design strategies, but share a common goal which is "the wellness of humans in accordance with environment"; and we may call this new trend as the second generation of healthy home environment. With no doubt, the concept of this current movement of home design is more advanced one, since they begin to embrace those factors neglected by the first generation: caring of non-standard humans, the increased sensibility in automation system, coping with the life cycle of home, and preservation of natural environment. It is certain that all these concerns are very timely since they would become increasingly crucial factors in the years to come. What lacks in them, however, is the fundamental concern about the inherent socio-spatial dimension of the house and its implication on the way we live; all of those issues treat the architectural space merely as an inert entity that simply follows functional requirements. Even the sustainable design, which includes this agenda in its broad meaning of the term, limits itself by vaguely suggesting the possibilities of spatial transformation of the house. In a sense, all these current issues still maintain the attitude of the first generation that a more precisely programmed house can guarantee a better life.

Considering above situations, the main theme of this research has sprouted: the freedom of the body, or rather bodily choices in domestic life, which has been half-forgotten during the course of housing evolution. It is not about the old concern for the flexible plan or the transformable house, but about the role of spatial settings in housing to liberate the bodily freedom to make a home interpretable. In a broad sense, it is about the anthropology of the human body in relation to the sociology of built environment. In what follows, it is argued that this issue presumably would emerge as the central point in the third generation of healthy home environment.

2. Polyvalency and the Levels of Living Frames

The issue of flexibility or adaptability is not new; it has constantly discussed in practice and education throughout the twentieth century. On a small scale, it could be applied to a single house level by providing movable partitions as in Schröder house, and on a bigger scale, to a mass construction level by utilizing the frame/infill concept as represented by SAR method. My definition of adaptability, however, is a bit different.

Architects have known for decades about another mode of adaptability, since Dutch architect Herman Hertzberger first used the term *polyvalency* in 1962. It means that a building can be used in different ways without adjustment to the way it is built [3]; thus it can be a more viable architectural feature than the movable partition in Schröder house or infill structure in SAR method when exposed to a long passage of time. Austra-

lian architect, Stefan Picusa clarified the concept of polyvalency by contrasting two kinds of adaptabilities, namely *inherent adaptability* and *potential adaptability* [4]. According to him, the former is "built into the initial design, giving the occupant choice through intentional ambiguity, within fixed physical constraints of a given plan" while the latter is achieved by "technically providing flexible spatial features like verandahs, undercrofts, roof spaces, and, more recently, demountable partitions and movable fittings" [4]. By inherent adaptability, he expressed exactly the concept, polyvalency. In the last century, however, it has received far less attention than potential adaptability [5], probably due to the modern functionalist spirit that everything, including adaptability, should be designed by the mighty hands of architects. For them, polyvalency in architectural space must have been felt as passive, neutral, and arbitrary. In the new millennia, however, its importance is slowly spreading out.

The term, "polyvalent" as an adjective, or "polyvalency" as a noun, would be used throughout this paper. However, it is not used simply at the level of architectural space but at the other levels including furniture and entire building as a block. Therefore, I will label any built structure as polyvalent if it could be utilized in more than a single way based on users' interpretations.

John Habraken distinguished five hierarchies of built environment: road network, building, partitioning, furniture, and body and utensil [6]. These levels represent a hierarchical enclosure system wherehigher levels regulate and affect the form of lower levels; thus, any analysis on one level cannot stand alone without the others. In this research, they would be reduced to four: building, partitioning, furniture, and body. If we evaluate these four levels in terms of speed, they would be ranked as in **Table 1**.

Above time scales and speeds are symbolic and relative idea, but it is evident that the higher the level, the lower the speed. As the building and the body are located in two extreme ends of the order, it seems natural that there is no way for the static building to catch up with the changing needs of the body through the passage of time; and this could be more so in the case of standardized

Table 1. Speed of the four levels.

Level	Speed
1	Buildings can stay on its site for more than 100 years
2	Partitioning, if there is no urgent need for change, more than 10 years
3	Furniture, if properly positioned, more than 1 year
4	Body, in any circumstances, less than 1 minute
Speed rank	Building < Partitioning < Furniture < Body

multi-unit housing which blindly attempts to accommodate a wide spectrum of lives in a small number of unit types.

Therefore, the main argument of this research would be based on this point: any pre-determined, or rather pre-fixed program of built structure cannot cope with the changing needs of human behavior, and thus, at some point, has to restrict the freedom of the body. In what follows, two of the above levels, those of furniture and partitioning, are discussed in detail. An argument would be made that modern furniture restricts the freedom of the body posture, and the programmed partitioning restricts the movement of furniture or rather spatial function. These two, combined, restricts the freedom of the body.

3. Programmed Partitioning and Human Movement

Looking at the old precedents of housing around the world, one can find that the rooms are generally for multi-purposes, and activities in a room can be transferred to other rooms without much conflict. The Palladian villa in **Figure 3** shows how this was possible. Rooms have plural number of access openings that lead to other rooms, not to a corridor or a central hall. When a row of rooms are directly connected sequentially like this, it makes a spatial configuration known as a "room-to-room enfilade". It is suggested here that this enfilade is an effective spatial device that can generate an enhanced degree of flexibility in space use as in Palladian villas. When two rooms of similar sizes are placed next to each other and directly accessed, they could support each other by accommodating similar activities when needed. When three rooms of comparable sizes are directly attached and accessed in a row, the room in the middle can support the two in each end. In this case, due to its innate ambiguity, this room can have a higher degree of adaptability; it can be a central zone that integrates the three or act as a mere buffering zone. This is how the enfilade works for polyvalency; since a room can readily support the adjacent one, activities are interchangeable. In this circumstance, a resident can perform the same activity in a different position; and thus his or her body is free from designated functions in space, and so does furniture.

If the room-to-room enfilade makes a ring-shaped spatial structure that can allow an unending circular movement, then the freedom of movement and the freedom of activity allocation is maximized as in Palladian villas; it generates a strong degree of polyvalency in the house. In the modern house, however, the ring structure is hard to be realized. This is because the modern home is becoming a container of an ever growing number of furniture that needs to be in touch with wall surface, and the ring structure requires at least two openings for each room by

sacrificing wall surface. Nevertheless, as long as rooms can allow it, the ring shaped spatial structure could be another strong device to induce polyvalency.

The use of the enfilade in the plan can be regarded as one of the defining characteristics of private houses. Gilles Barbey noted that the difference between the residential housing and the institutional housing, such as prison, dormitory, and monastery, can be seen by the dominant use of the enfilade in the former and the corridor in the latter for the distribution of space [7]. Considering the fact that the residential housing needs a higher degree of adaptability than the institutional housing, in order to accommodate a wide spectrum of living patterns and their changing needs in time, this seems a natural design solution. In the twentieth century, however, we have observed that more and more houses adopt the corridor for the flow distribution, unlike their old counterparts. What could be the consequences of this transformation in these modern houses?

In the seventeenth century UK, there appeared a passage in the middle of the house plan as in **Figure 1**. By using this new spatial device, a direct access to a room from the public zone, without passing through other private rooms, was made possible. Followingly, access routes in the house were made simple and easy with improved privacy, while sacrificing the route choice by the users. According to Robin Evans, the evolution of the house plan towards a more privacy-oriented arrangement by means of the corridor is to facilitate "purposeful communication" while reducing "incidental communication" between rooms and residents [8]. Yet, in the light of polyvalecy, "incidental" could be more valued than "purposeful"; it is not the "programmed" route but the "incidental" route that can generate a new possibility of spatial use—a passive way of increasing the freedom of the body.

The use of the corridor certainly diminishes the interchangeability of activities that "requires a wide range of possible connections for services" [9]. In other words, the modern plan layout where a multiple number of rooms are exclusively accessed from a single room, *i.e.* the corridor or a central hall, cannot support the polycalency in life, because this type of plan is "less capable of being adapted to suit different living patterns" [3].

As described above, there is an evident tendency that the room-to-room enfilade enhances activity interchangeability, while the corridor does not. Based on this line of thought, it may be said that the degree of polyvalency can be measured, to a certain degree, by investigating the usage of the corridor in a given plan. **Figure 2** shows the urban traditional house in Seoul, Korea that prevailed between the 1930s and 1950s. Here, the central courtyard directly links other rooms, making the enfilade without corridors, and also *maru*—a space with raised wooden

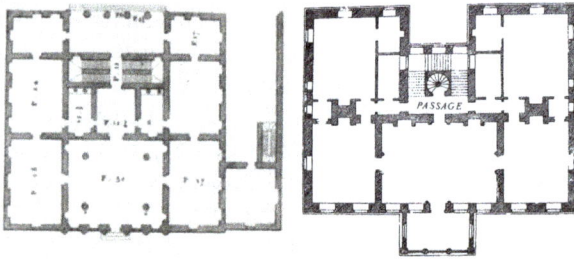

Figure 1. Palazzo Antonini (1556, left) by Andrea Palladio and Amesbury house (1661) by John Webb.

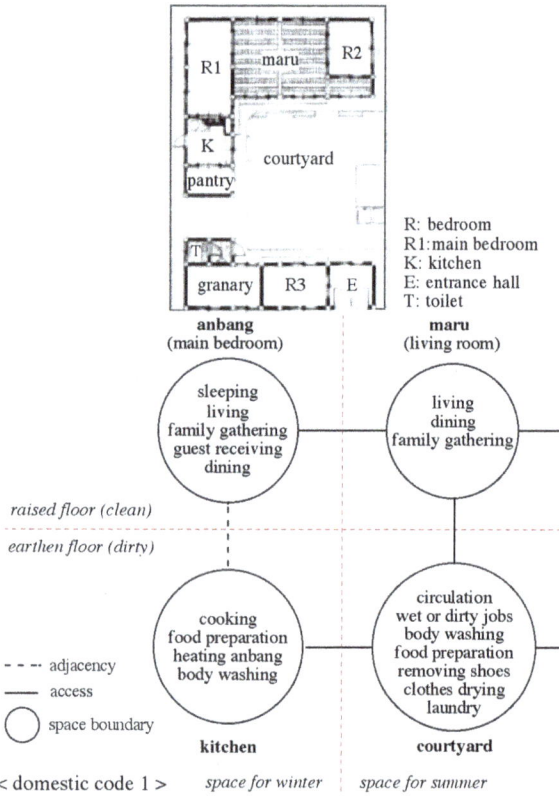

Figure 2. Urban traditional house and its activity diagram [10].

floor—offers the enfilade approach to other two bedrooms on both sides. On the right side is the diagram that shows the four spaces—anbang, maru, courtyard, and kitchen—linked together, and the activities executed in the rooms.

It is found that anbang and maru support each other with the functions of living, dining, and family gathering, and kitchen and courtyard with those of food preparation and body washing. Without corridors or other transient space for circulation, the enfilade structure of this house could offer the freedom of living in relation to time, seasons, and personal choices.

Figure 3 shows the apartment house plan in Seoul in the 90s. As all the circulation movements are concentrated in the central hall, which is marked H in the plan,

this space performs the role of a corridor. This hall space can be seen, at first sight, as a part of the living room, since it is open to it without partitions. However, it is evident that its function as a distribution core clearly distinguishes this space from the living room zone. The diagram on the right also confirms this fact. The central hall links all the access relations from the middle of the house; therefore, residents have to go through this traffic center to move to other spaces in the house. In this type of house plan, it may be hard for a room to support the activities of the other room as in the urban traditional house.

During the transformation process of the apartment houses, there were plans that still preserve the room-to-room enfilade as shown in **Figure 4**.

In these plans the living room and the anbang were directly connected, and thus each could support the other. As the number of furniture grows in the house, the characteristics of each space were defined more specifically.

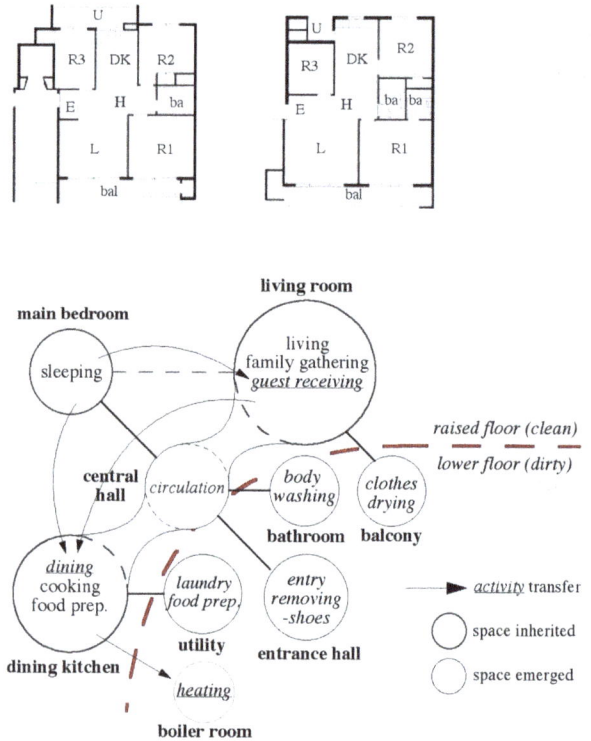

Figure 3. Typical apartment house plan in the 1990s [10].

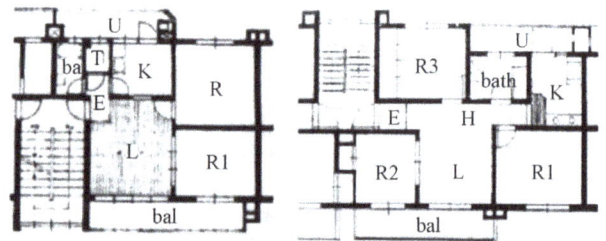

Figure 4. Earlier apartment house plans in the 1960s.

Therefore, each room had to be more independent, performing specialized functions, and it needed more wall surfaces. All these change demanded, as a convenient solution, the central corridor as a powerful integrator of circulation.

In sum, Korean houses have evolved from the enfilade to the corridor; from interchangeability to fixed function; from a personal choice of living to a pre-determined living. The corridor certainly reduced the possibility of polyvalence, and even those early apartment plans that still preserved the enfilade gradually disappeared.

This observation can be further corroborated by investigating multi-unit houses with different size. **Figure 5** shows five staircase access plans with different number of bedrooms and their corresponding graph representations. They are the most widely used plans in Metropolitan Seoul between 1962 and 1990 according to Kim and Park's study [11]. Pointing out that in Seoul a small number of typical plans are repeatedly produced, they argued that these plans are designed in the same manner regardless of their sizes. Below each plan is the graph developed by this author to effectively show the access and adjacency relations of rooms. In the graph, rooms are represented by bubbles, and access and adjacency relations between them are represented by continuous and dotted lines respectively. The gray rectangular box means the interior area and thus balconies or utilities are generally positioned outside of it. This graph-theoretic representation, thus, can explain both geometrical and topological properties of the plan in a single standardized format.

What these graphs reveal is that all the plans of different sizes have one prominent feature; most of the rooms are connected to or through the central hall which is marked "H". Now, no enfilade and no ring structure

can be seen in the people's house in Seoul, and the chance to experience the possibility of bodily freedom has disappeared.

4. Discussion and Conclusions: Polyvalent Houses

What is suggested in this section is the theoretical ways of making polyvalent houses. Some good examples are analyzed to find out their spatial logic. **Table 2** shows two different ways of making enfilades in the house. A house needs to have service spaces which can make a core. When the core is placed in the middle, the enfilade is made that will automatically form a ring. Another way is to put the split cores on each side in order to provide a central hall that is connected to other rooms in an enfilade ways. This, however, do not provide rings. When enfilades are combined with rings, their effects are multiplied. Dapperbuurt house in **Table 3** shows how this simple house can have 14 different sets of routes by closing or opening the movable partitions or doors. Compared to this, Pieter Vlamingstraat house at the bottom of **Table 3** has only three variations in the route choices due to the position of the core on each side; the cores block the ringy structure of space.

To make a ring in the house, it is necessary that at least one partitioned space has two access points or two doors. This, however, is a local structure that does not have a big impact to free the functional fixation of the house. It is better to have a global ring that connects rooms in the periphery of the house. The two basic ways in **Table 2** show why central core is more effective in creating enfilades and rings. However, the central hall type can also be made to provide rings as well as enfilades.

In **Table 4**, the plan above makes a ring around the

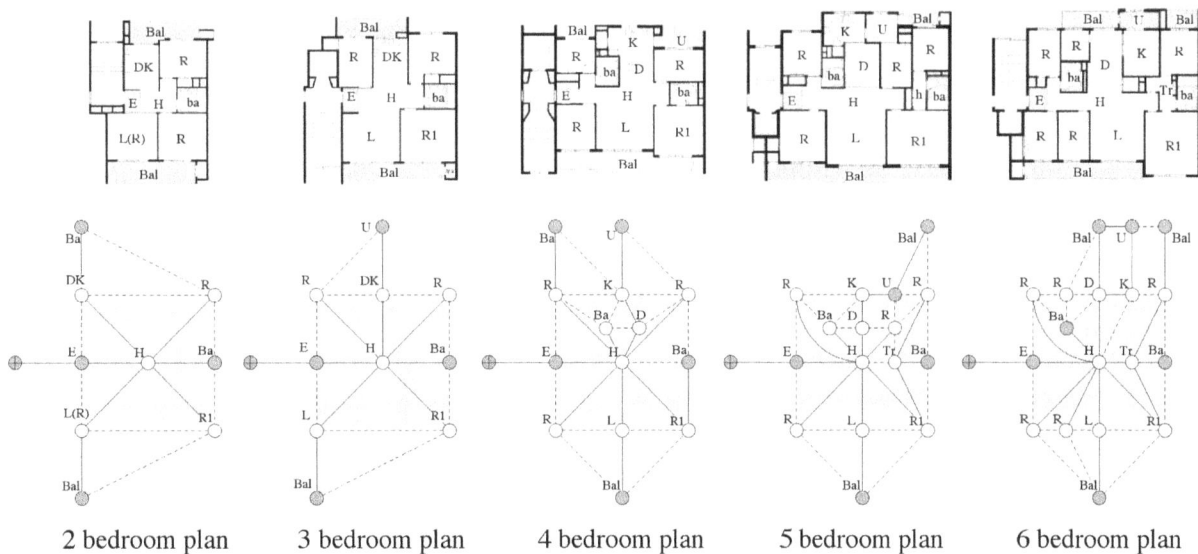

| 2 bedroom plan | 3 bedroom plan | 4 bedroom plan | 5 bedroom plan | 6 bedroom plan |

Figure 5. The most popular staircase type plans in Seoul between 1962-1990 [12].

Table 2. Two Basic ways of making enfilades [13].

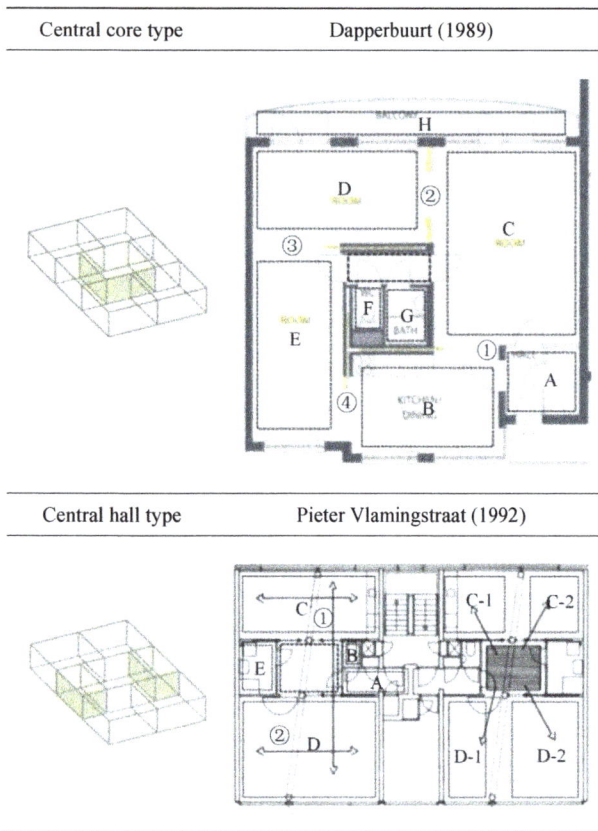

Central core type	Dapperbuurt (1989)

Central hall type	Pieter Vlamingstraat (1992)

Table 3. Comparing possible routes between two types.

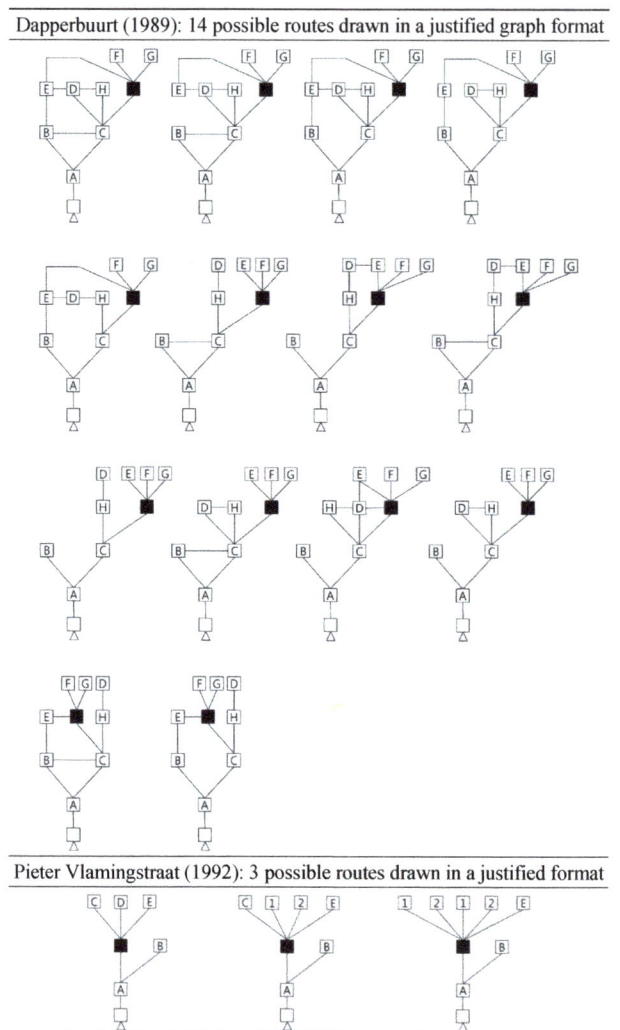

Dapperbuurt (1989): 14 possible routes drawn in a justified graph format

Pieter Vlamingstraat (1992): 3 possible routes drawn in a justified format

Table 4. Two alternative ways of making rings [14].

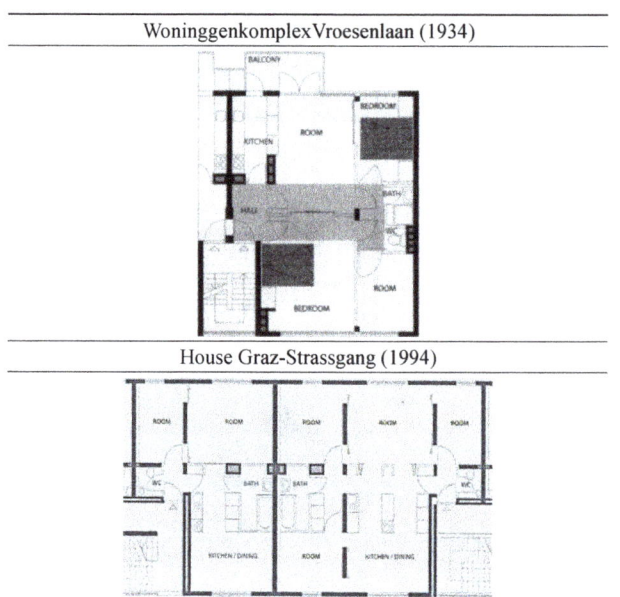

Woninggenkomplex Vroesenlaan (1934)

House Graz-Strassgang (1994)

central hall by placing two doors on each side and sliding partitions between them. In the plan below, central hall is sub-divided by service installations and discontinuous walls within the unit. This plan thus provides many local rings as well as global rings.

This research attempted to illuminate the forgotten usability of the inherent adaptability which is against the tight-fit space programming of the modern functionalists who insist on the "form follows function" attitude.

It is when this deterministic characteristic of domestic space is dissolved that the pre-condition for the freedom of the body, which allows more fluid and contingent functional possibilities, could be established.

We can choose our own route in urban settings but not in modern domestic settings. Through the process of housing evolution, we have lost, consciously and unconsciously, the freedom of bodily move and the freedom of bodily posture. Some propositions can be made for the recovery of bodily freedom in the house as in **Table 5**.

These five hierarchies of remedy reflect the levels in architecture, from a body to a building in the increasing order. If these strategies are more carefully evaluated for the design of housing, then we can build a healthy home in the deepest meaning of it. The healthy housing design is the one that encourages the various type of bodily movement, without restricting it in the name of move-

Table 5. Hierarchies of remedies for bodily freedom in the house.

	Hierarchies of Remedy	Hierarchies of Bodily Freedom
1	Recognition of postural potentials	Freedom of using muscles
2	Re-evaluation of "sitting furniture" shapes	Freedom of making postures
3	Renovating the power supply installations	Freedom of locating wired duties
4	Re-thinking of function-fixed rooms	Freedom of distributing activities
5	Re-evaluation of building circulation	Freedom of selecting routes in buildings

ment optimization and functional convenience.

5. Acknowledgements

This research was supported by Basic Science Research Program through the National Research Foundation of Korea (NRF) funded by the Ministry of Education, Science and Technology (NRF-2011-0012359).

REFERENCES

[1] A. Ravetz, "The Place of Home: English Domestic Environments, 1914-2000," E & FN Spon, London, 1995, pp. 166-167.

[2] S. Muthesius and M. Richardson, "Continuity and Change," The Transformable House, John Wiley & Sons Limited, Hoboken, 2000.

[3] B. Leupen, "Towards Time-Based Architecture," 010 Publishers, Rotterdam, 2005.

[4] S. Pikusa, "Adaptability," *Architecture Australia*, Vol. 72, No. 1, 1983, pp. 62-67.

[5] R. J. Lawrence, "Housing, Dwellings and Homes: Design Theory, Research and Practice," John Wiley & Sons Limited, Hoboken, 1987.

[6] N. J. Habraken, "The Structure of the Ordinary," MIT Press, Cambridge, 1998.

[7] G. Barbey, "Social Space in Residential Environments: the Importance of Spatial Archetypes and Their Implication for Design," Progetto Finalizzato Edilizia, Milano, 1992.

[8] R. Evans, "Translations from Drawing to Building and Other Essays," Architectural Association Publications, London, 1997.

[9] F. Bijdendijk, "Solids," In: B. Leupen, Ed., *Time Based Architecture*, 010 Publishers, Rotterdam, 2005.

[10] K. W. Seo, "The Law of Conservation of Activities in Domestic Space," *Journal of Asian Architecture and Building Engineering*, Vol. 5, No. 1, 2006, pp. 21-28.

[11] S. Kim and Y. Park, "Inflexible Pattern of Apartment Unit Plan: The Comparative Analysis of the Public and Private Sector Apartment-III," *Journal of Architectural Institute of Korea*, Vol. 8, No. 7, 1992, pp. 73-83.

[12] K. W. Seo, "Space Puzzle in a Concrete Box: Finding Design Competence That Generates the Modern Apartment Houses in Seoul," *Environment and Planning B: Planning and Design*, Vol. 34, No. 6, 2007, pp. 1071-1084.

[13] B. Leupen and H. Mooij, "Housing Design: A Manual," NAi Publishers, Rotterdam, 2012.

[14] T. Schneider and J. Till, "Flexible Housing," Architectural Press, Oxford, 2007.

12

Korean Academic Librarians' Recognition of the High Density Book Storage System

Joonsuk Ahn

Department of Architecture, Kyonggi University, Suwon, South Korea.

ABSTRACT

Korean academic libraries are facing a serious space shortage problem due to the inability to uphold the rapidly increasing amount of printed materials despite having expanded the number of physical facilities. Data computerization has been considered as a solution to the issue, but deliberation for the High Density Book Storage System has been on the rise because of its impressive method of preserving printed materials in a realistic facility. Despite the different methods of print material storage, Korean academic libraries have largely focused on investing in the least efficient method of compact shelving to solve this issue. It is hypothesized that the misuse of funds on inefficient systems is occurring due to the lack of knowledge about the high-density book storage systems like the Harvard model. In order to propose a realistic solution to the academic library space shortage crisis on a logical basis, it is imperative that a study of academic librarians is conducted to investigate their knowledge on such efficient storage systems.

Keywords: Space Shortage Problem; Academic Library Facility; High Density Book Storage; Harvard Model

1. Introduction

Korea's rapid economic grown in the 1980s brought enormous spatial expansion to the physical facilities associated with academic libraries. The growth allowed for a new era in establishment of modern Korean academic libraries and fostered a positive impact on collegiate environments nationwide. Beginning in 1955 with only 43 total public and private academic libraries in the entire country, Korea reached a total of 523 facilities in 2009.

More importantly, there was a noticeable increase in book quantity. With 1,297,034 books observed in Korean academic libraries in 1955, the number was registered at 121,479,083 by 2009 [1]. **Figure 1** presents the rapid growth curve of printed materials observed in academic libraries from year 1955 to 2009. **Table 1** shows the comparison of number of books, academic library facilities, and librarians between 1955 and 2009 in Korea. The total of amount of books had expanded by 9370%, and that increase was almost 8 times faster than the growth of academic library facilities. This exponential growth created a serious space shortage problem for all Korean academic libraries and slowly led to academic environ-ment degradation. **Figure 2** represents the problem graphically.

The vastly increasing quantities of printed materials and the library space shortages brought about by it are the biggest problems facing current academic libraries in Korea. Korean academic libraries have been in search of efficient book storage systems to solve the issue. The movable compact shelving unit, also referred to as the 'mobile rack', is widely utilized. Since the adoption of the movable compact shelving system, open access systems have been put in place. The open access system provides higher service quality for its users, but it is limited in space efficiency. The alternative closed access system lacks a user-friendly operating system; however, it has much higher space efficiency at a significantly lower cost. Judging by the cost-benefit tradeoff, the closed access high-density book storage system is the only logical option to resolve the academic libraries' space shortage crisis [2].

The study was projected to assess Korean librarians' understanding of high-density book storage facilities used for academic and research purposes, and identify

Table 1. Number of academic library facilities, books, and librarians in 1955/2009.

Year	1955	2009
Academic Libraries	43	523
Books	1,297,034	121,479,083
Librarians	207	3686

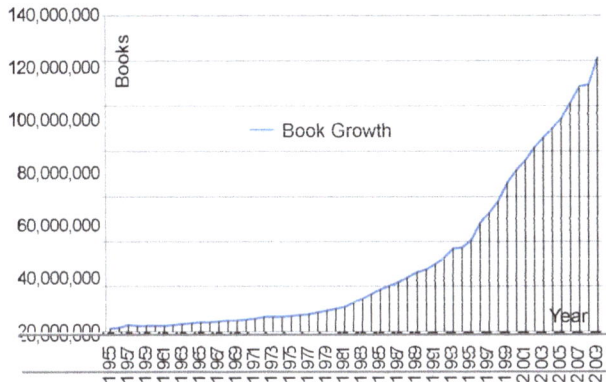

Figure 1. Book accumulation of academic libraries in Korea.

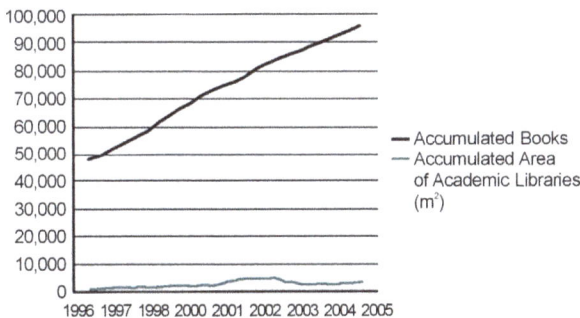

Figure 2. Comparison of book growth to library facility growth [3].

reasons for the lack of establishment of such facilities in Korea. The analysis of the survey responses will serve as a method to further resolve the space shortage crisis in a workable fashion.

2. High Density Book Storage: Harvard Model

In the 90's, Harvard began its construction of high density book storages, the "Harvard Model", as an efficient method of preserving low use print materials to serve as the solution for the space shortage problems in academic libraries. The Harvard Model provides extremely high space efficiency at a low cost. The original idea for this system was inspired by the distribution and warehouse industry. This system has spread all over the world and has now become a development standard for book stor-

age facilities [4]. **Figure 3** shows the shelving system of Harvard Model at Rice University Library Service Center. More than 100 high-density book storage facilities have been built worldwide, 73 of those located just in the United States.

However, it has been found that Korean academic librarians have a definite preference for open access systems and there have not been any Harvard model storages constructed in Korea. An overwhelming majority of librarians continue to prefer expensive library buildings in lieu of low cost storage facilities despite the fact that new open access library facilities will never resolve space shortage problems. Many libraries have installed movable compact shelving units or remodeled their facilities to further accommodate, but these methods only postpone the impacts of the space shortage problem. These libraries inevitably face the same space shortage problems just after a couple of years after the completion of such constructions [5].

3. Survey Questionnaires & Evaluation

Previous surveys show that Korean academic librarians have acknowledged the need for high-density book storage systems in order to resolve space shortage problems. It is proposed that these librarians continue to solely utilize the movable compact shelving system, the mobile rack, with the exception of Sungkyul University's ASRS (Automated storage and retrieval system), because of their lack of understanding of high density book storage systems to make an informed decision on selecting the most efficient storage methods. As librarians are the most important individuals when making decisions on new constructions, remodeling projects, and operation systems of their respective academic libraries, it is important to investigate librarians' understanding of the high density book storage systems and their knowledge of opportunities to resolve space shortage problems.

Figure 3. Harvard model storage system at Rice University.

Questionnaires regarding the issue above were distributed to 463 academic librarians through email. Replies were received through "Google Drive" from 75 librarians at 169 universities in Korea from June 10-16, 2011. The same questionnaires were sent to another 1186 academic librarians at accredited 4-year universities resulting in 182 replies throughout a period from June 26th through June 30th, 2011. The questionnaires resulted in a response rate of 15.5% and a total of 257 responses.

The survey was comprised by 7 questionnaires as below:
- The necessity for the adoption of high density book storage systems to resolve the space shortage problem
- Knowledge about high density book storage types and operational options
- Critical decision making elements on library facility development
- Preferred methods of obtaining extra space to preserve printed materials
- Application plans for any possible available space
- Plans of developing extra book storage space with the exception of building a new library
- Understanding of cooperative book storage facilities

3.1. How Effective Do You Believe the High Density Book Storage System Will Be in Reducing the Space Shortage in Your Library?

The purpose of this question was to measure the librarians' opinion on the efficacy of high-density book storage systems in solving the space shortage issue. **Figure 4** represents the replies of this questionnaire. Out of 243 total replies, 159 (53%) and 100 (41%) responded that the high-density storage system would be very effective and effective, respectively. Judging from the data that shows a large majority, 94% of the responses, were positive for the implementation of high-density storage systems, it is concluded that most academic librarians are in favor of introducing this type of method into the nation's library system. From the small percentage of negative responses (1%), it can be said that there are hardly any opinions opposing high-density book storage systems. Those librarians who responded with a negative attitude towards this type of management system showed a lack of understanding of such systems and an extreme preference for open access management.

3.2. Choose All High Density Book Storage Types in Which You Are Familiar with the Method of Operation

This question was posed to librarians so that they would choose all types of high-density book storage systems in which they understood all facility and managing systems

in order to investigate their level of understanding for each type of system. The survey result is shown in **Figure 5**. As expected, the compact shelving system (mobile-rack) received a large sum of 221 votes (91%) followed by the Automated Storage and Retrieval System (ASRS) with 93 votes (38%); however, only 4% of librarians identified as fully understanding the Harvard Model, in addition to a surprising 3% of librarians which stated that they had no understanding of high density book storages. It is inferred that Sungkyul University's 2010 construction of ASRS models helped in informing librarians about this specific method possibly resulting in the high number of votes for this system. It was unexpected that so many librarians, 34%, showed a high understanding for the outdated Multi-Tiered Stack Core System, but the result is interpreted as the librarians' informed knowledge about the history of librarians and their shelving systems.

The fact that Korean academic librarians have such limited knowledge about high-density book storages systems like the Harvard model, which has already become a standard in the United States and Europe, shows the librarians' lack of understanding is even far more limited than as previously predicted. The results of this survey show that it is imperative that further knowledge about this type of system is more widely distributed.

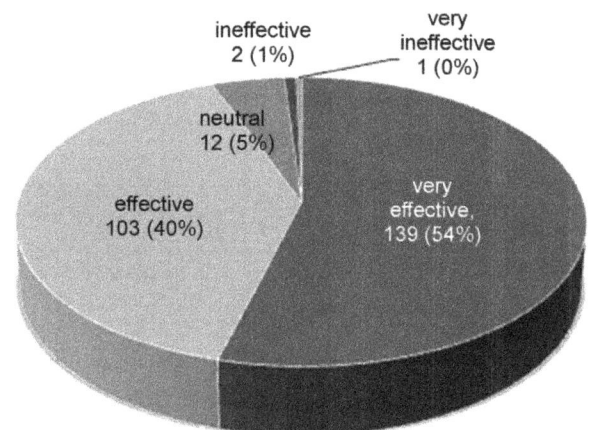

Figure 4. Necessity of high density book storage.

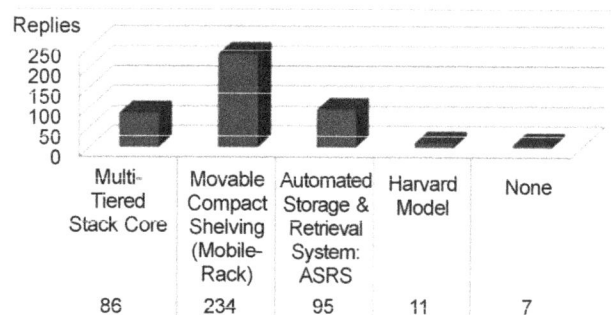

Figure 5. Awareness of high density book storage types.

3.3. What Is the Most Critical Element You Consider When Selecting a Book Storage Type?

The 66% of librarians replied that the book storage capacity per square footage, therefore space efficiency, was a critical element when selecting a storage type. This result is a strong representation of the magnitude of the academic library space shortage problem alongside the librarians' desire to resolve the issue. It signifies that spatial efficiency, rather than construction cost, should be deemed the highest priority when selecting a storage system. From the results, it can be inferred that librarians would prefer a facility with guaranteed space efficiency even with a trade off with time spent on budget collection compared to a shortsighted facility.

19% of the responses chose construction cost as seconding space efficiency in elements to be considered when choosing a storage facility. Difficulties in fulfilling a budget to construct a book storage system pushed opinions to prefer economical and practical facilities. A small minority of the responses chose options such as operational manpower (4%), operation and maintenance cost (5%), facility location (6%), showing that these other alternatives were far less critical compared to establishment cost and space efficiency. **Figure 6** shows the replies of this question. The Harvard Model is the most space efficient of the book storage systems with a low operation and maintenance cost needed for manpower in addition to low construction costs. Judging from the responses received from the pool of librarians surveyed, such systems that fully accommodate for all the considerations are the best options that should be introduced into the nation.

3.4. Which Option Is Best for Securing Extra Library Space? (Under Limited Budget)

Responses shown in **Figure 7** indicated that 40% of librarians' preferred storage options that are economical and better insure security of budget. However, large opinions showed that librarians still largely believed that at equal costs, they preferred open access to closed access services even if that meant less space efficiency (37%). It is thought that this is because when the questionnaire was formulated, the survey did not mention that open access storage systems preserved a mere 10% of what a high density book storage would under the same given square footage. Because librarians lack full understanding of the space efficiency potentials of the high-density book storage system, they still select the open access storage that matches the traditional library structure at a high percentage. Regardless of construction costs, librarians who supported the open access storage systems (15%) were relatively higher than those who

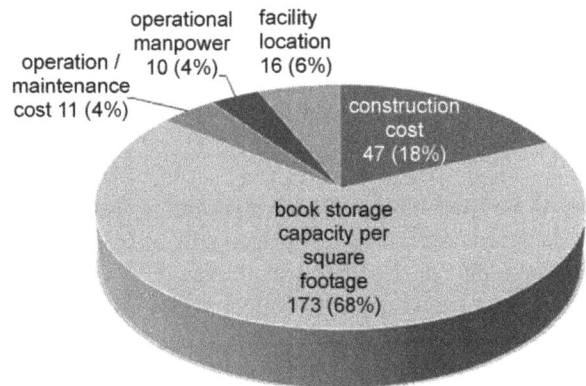

Figure 6. Critical decision element on storage type selection.

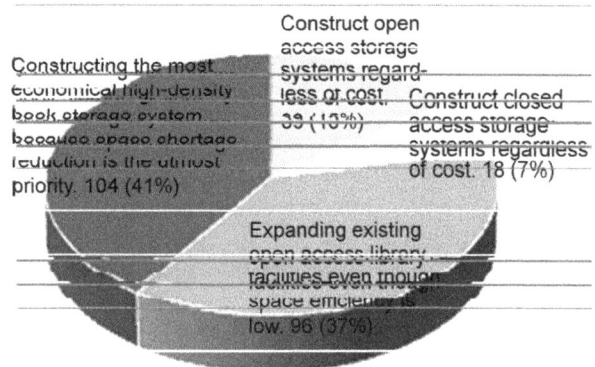

Figure 7. Options for securing extra space for library.

supported the high density storage systems (9%), resulting in an overall 52% of votes preferring the open access system compared to the 47% that preferred the high density storage. It is inferred that the preference for the less efficient system is due to the insufficient understanding of the fairly new concept of high-density book storages. In addition, it can also be inferred that the librarians and other library staff that are under the constant stress of space shortage unquestioningly prefer the open access service because they aren't completely aware of the full import of the issue.

3.5. If It Were Possible to Transfer 500,000 Books to a New High Density Book Storage, What Would Be the Biggest Benefit to Your Library?

Improvements to shelving arrangements and operational convenience (52%) and the addition of a rest area to improve environmental quality (20%) were the top two potential usages of the new available space acquired from the implementation of the high-density book storage system. The opinions of librarians that believed that benefits brought about by an information commons and learning commons (12%) were important showed new up and coming trend of considerations for newly available space followed by the suggestion to add more reading space and open access shelving (10%), and adding space for new equipment for academic use (5%). **Figure 8** represents the librarians' opinion on the new available space usage.

Over half of the opinions stating that extra space should be used for better shelving arrangement provide evidence of the librarians' strong will to improve the spatial quality of the library. Furthermore, the other inclination to use space to add more resting areas shows the movement of librarians' ideology of "Library as a Place" [6], showing that economic growth naturally coincides with cultural development.

3.6. If You Could Not Secure a Budget for an Independent Library Facility, Which Do You Believe Is the Most Practical Plan among the Options Listed below?

For alternatives for obtaining book storage space, extension of closed access book storages received 39% of the votes, being the most preferred, followed by utilization of other existing facility (storage, basement, etc.) on campus at 29%, open access book storage extension with 13%; building rent for remote off campus book storages at 3% was found as a minority opinion. **Figure 9** shows the result of this questionnaire survey.

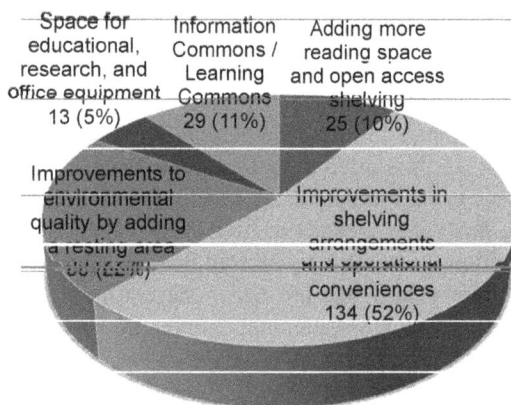

Figure 8. Potential usage of extra space provided by high density book storages.

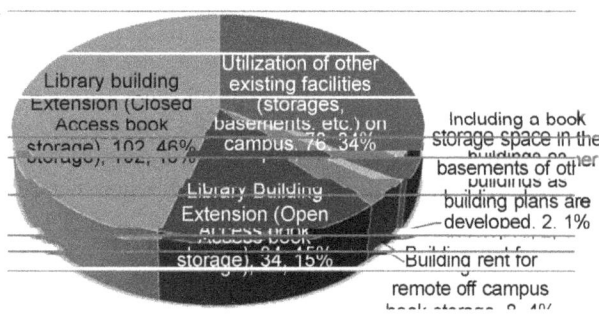

Figure 9. Alternatives to high density book storage.

Librarians showed no preference in including a book storage space in basements of developing buildings, albeit the option would allow for the implementation of a practical plan for maximum usage of real estate while providing a properly designed environment for book storage. It is probable that the idea of a basement brought about an image of a dark and humid environment with deficient ventilation that was negatively perceived as a good space for book storage; it is also possible that the librarians were displeased with the idea of storing books in a location other than a library.

Misconceived by those librarians, this alternative does not conceptualize an environment in which an already existing low quality basement is transformed into book storage. Rather, the option would allow for a new library facility to be constructed underground. If moisture control and appropriate ventilation were implemented, the benefits of an underground facility, including heat and sound insulation quality, protection from direct sunlight, and structural stability, would serve remarkably as book storage. The benefits of a basement facility are currently greatly underappreciated, thus resulting in this particular survey result.

3.7. What Do You Believe Is the Best Way to Accomplish a Cooperative Storage System in That Universities Come Together to Construct Book Storage under an Economical Budget?

Table 2 indicates the librarians' preference of the cooperative book storage. Librarians most preferred the cooperative book storage system that implemented joint ownership, joint management (35%). The options of preserving books together with separate ownership and separate management of a shared storage were voted with similar preferences (23% and 20%, respectively). Shared storages being built by the university with funding, but renting out the facility and having the cooperative storage stores and manages its own books without duplicates both resulted in 10% of the votes. Unique from the other surveys given thus far, this particular questionnaire presented with a tendency to present preferences for all options fairly consistently.

However, as a result of government oversight that induces extreme competitions amongst Korean universities, the goal of joint preservation, joint ownership, and joint management among these educational institutions will be a difficult target to meet. In consequence, joint construction of book storage with independent management will be the most realistic goal for Korean institutions.

4. Conclusion and Discussion

As concluded from the results of the full survey, the

Table 2. Preference of cooperative book storage types.

Cooperative Book Storage Types	No. of Replies (%)
Under agreement among all the institutions involved, establish a cooperative storage system under joint ownership and joint management and work to share data to preserve only single copies of printed material.	90 (35)
The cooperative storage owns and manages the facility and books independently, preserving single copies of all printed material	25 (10)
Under full agreement by the involved institutions, books are preserved together, but owned separately by respective universities	57 (23)
The storage is shared, but managed separately	56 (22)
Book storage is built by institution with funding and rented or co-managed by other institutions	25 (10)
Total	253 (100)

disuse of high density book facilities such as the Harvard Model that compose of 60% of American libraries and the overwhelming use of the movable compact shelving (mobile-rack) system by all Korean universities (with the exception of Sungkul University) signify Korean academic librarians' limited knowledge of high density book storage types and storage alternatives.

Considering the outstanding space efficiency and economic feasibility of high-density book storage systems, the questionnaires were returned with somewhat unexpected replies; however, such responses may be rationalized if the librarians answered the surveys with the thought of movable compact shelving in mind because of their lack of understanding of the high-density book storage types. Judging from the inadequate understanding of high-density book storage systems by librarians who are considered experts in the field of library management, it can be said that the public's awareness of such facilities is even more minimal.

Nevertheless, the librarians unanimously adhered to the idea that resolving space shortage problems was their primary priority and in order to meet that goal, facilities must have the outstanding space efficiency. Informing these experts with the strengths and weaknesses of vari-

ous book storage types and then reconducting the surveys will ultimately result in meaningful changes to responses. Because there is no record of cooperative storage precedents in Korea, the responses regarding such facilities are seen to have resulted in more notional responses; however, it can be understood that work experience has naturally rooted understanding in our library experts.

This study was conducted to gauge Korean librarians' understanding of high-density book storage facilities widely used for academic and research purposes in highly developed countries and identify reasons why such storage systems were not constructed in Korea. The questionnaire serves as a measure of awareness of librarians on the library space shortage crisis. By analyzing the data retrieved from this survey, we can further work to resolve the space problem in a practical yet meaningful manner.

REFERENCES

[1] Korean Library Association, "2010 Korean Library Year-book," Korean Library Association, Seoul, 2010

[2] J. Ahn, "Space Efficiency Improvement Plan for Academic Library Storage: Focusing on High-Density Storage Facilities" Ph.D. Thesis, Korea University, Seoul, 2011.

[3] H.-J. Yi and Y.-J. Nam, "A Study on the Space Rebuilding Using the Digital Preservation," *Proceedings of the 13th Annual International Conference of the Korean Association of Information Management*, Seoul, August 2006, pp. 201-202

[4] J. Ahn, "Study on High-Density Library Storage as a Solution to the Space Shortage Problem: A Case Study of Rice University Library Service Center," *Journal of Korean Institute of Educational Facilities*, Vol. 17, No. 6, 2010, pp. 23-30.

[5] L. Payne, "Library Storage Facilities and the Future of Print Collections in North America," OCLA Online Computer Library Center, Inc., Dublin, 2007

[6] K. Antell and D. Engel, "Stimulating Space, Serendipitous Space: Library as Place in the Life of the Scholar," In: J. E. Buschman and G. J. Leckie, Eds., *The Library as Place*, Libraries Unlimited, Inc., Westport, 2007, pp. 168-176.

13

A Multi-Criteria Decision Support System for the Selection of Low-Cost Green Building Materials and Components

Junli Yang, Ibuchim Cyril B. Ogunkah*

Department of Construction, School of Architecture and the Built Environment, University of Westminster, London, UK.

ABSTRACT

The necessity of having an effective computer-aided decision support system in the housing construction industry is rapidly growing alongside the demand for green buildings and green building products. Identifying and defining financially viable low-cost green building materials and components, just like selecting them, is a crucial exercise in subjectivity. With so many variables to consider, the task of evaluating such products can be complex and discouraging. Moreover, the existing mode for selecting and managing, often very large information associated with their impacts constrains decision-makers to perform a trade-off analysis that does not necessarily guarantee the most environmentally preferable material. This paper introduces the development of a multi-criteria decision support system (DSS) aimed at improving the understanding of the principles of best practices associated with the impacts of low-cost green building materials and components. The DSS presented in this paper is to provide designers with useful and explicit information that will aid informed decision-making in their choice of materials for low-cost green residential housing projects. The prototype MSDSS is developed using macro-in-excel, which is a fairly recent database management technique used for integrating data from multiple, often very large databases and other information sources. This model consists of a database to store different types of low-cost green materials with their corresponding attributes and performance characteristics. The DSS design is illustrated with particular emphasis on the development of the material selection data schema, and application of the Analytical Hierarchy Process (AHP) concept to a material selection problem. Details of the MSDSS model are also discussed including workflow of the data evaluation process. The prototype model has been developed with inputs elicited from domain experts and extensive literature review, and refined with feedback obtained from selected expert builder and developer companies. This paper further demonstrates the application of the prototype MSDSS for selecting the most appropriate low-cost green building material from among a list of several available options, and finally concludes the study with the associated potential benefits of the model to research and practice.

Keywords: Analytical Hierarchy Process (AHP); Decision Support System (DSS); Low-Cost Green Building Materials; Decision Analysis; Material Selection Factors

1. Introduction

As the green building movement begins to sweep through the housing construction industry, the application of cost effective and energy efficient building materials has become necessary in today's demanding economic market [1,2]. Recent discussions on the need to lower the growing demand for conventional sources of energy have highlighted the value of using low-cost green building materials and components, given their lower cost and energy requirements [3,4]. Evidence from previous studies has proven that implementing such products in construction has the potential to not only reduce health and environmental effects, but to also bring savings from energy, maintenance, and operational costs [5-9]. Yet, research has consistently shown that the patronage for such materials in housing construction is still at a very low level in comparison to many other conventional building materials [8,9]. Recent studies [10,11] argue that several attempts to adopt low-cost green building materials for housing design projects have generally been viewed as challenging, given that most designers are vaguely informed about the full life-cycle impacts of such products. They note that information relating to the impacts of such building materials in the housing construction sector appears to be less available, as evidence [11,12] indicates that only a small proportion of design and building professionals seem to have sufficient knowledge that could allow effective decision-making. Ashraf [11] and Zhou et al. [12,13] suggest that maximizing their potential use in the housing industry requires seamless access to appropriate informed information and full understanding of the various options available, so as to inform decision trade-offs at the design stage.

Despite the availability of accurate and reliable data, Seyfang [14] and Malanca [15] however, noted that most designers are found to make decisions regarding the selection of such materials on the basis of their past experience. They observed that inexperienced designers generally engage the traditional mode of selection, by relying on subjective individual perceptions of values and priorities in the material selection process, which rather than facilitate or drive their design ideas, appear to do the opposite thereby limiting creativity and sometimes resulting in considerable frustration [16,17].

Trusty [18-21] & Woolley [22] further disclosed that existing databases on such materials and their formats are not designed to efficiently and directly provide such information to decision makers. They note that the available data on such materials are normally in the read-only format, and are stored in various operational databases that are not easily accessible to decision makers in usable forms and formats. As a result, decision-making failures during the planning and design stage(s) of low-cost green housing projects hinder their use in terms of their industrial capacity utilisation in the housing industry.

While several studies [14,15,20] have emphasized the relative importance of information access in aiding well-considered and justifiable material choices during the early stages of the design process, Wastiels et al. [16] argue that the existing material selection method focuses mainly on limited aspects of such materials, in terms of their properties and factors that influence the decision-making process. Quinones [17] asserted that some low-cost green building materials, for example, contain high embodied energy that leads to ecological toxicity and fossil fuel depletion impacts during their manufacturing phase. She argued that ignoring the relevant factors or properties of any of such materials during the crucial material selection phase could reduce the effective life of that product to less than half of its normal effective life span.

Moreover, Seyfang [14] and Trusty [18-22] argue that choosing the right materials for a particular project can be a very complex decision-making task, given that the selection process is influenced and determined by numerous preconditions, decisions and considerations. They suggested the idea of a decision support system (DSS) as a useful aid in making quick and critical decisions during crucial material selection process. They stressed that the considered approach to encourage the wider scale use of low-cost green building materials in mainstream housing should enable design professionals to have easy access to adequate information on the available options, hence, making the selection results more reasonable and bringing more standardization to the material selection decision-making process at the design stage. They went on advising that whatever method is employed must be such that it allows comparison of not only the cost or technical performance of such materials, but also able to take into account several decision-making criteria, so as to derive conclusive and valid evidence of the differing impacts of various material alternatives.

While there seem to be no compelling evidence of technical research on a holistic approach used by design professionals for the evaluation and selection of building materials, previous material assessment models such as the Leadership in Energy and Environmental Design (LEED) and Building Research Establishment Environmental Assessment Methods (BREEAM), have shown great promise for guiding evaluations of material predictor performance [23]. The findings of the main research study yet, criticised and noted the flawed existing support systems for being partially objective and fraught with problems of fairness [24]. The study revealed that existing methods are found wanting in that they are culturally implicit, and that such methods or tools treat the sustainability [of the] wider built environment as simply a matter of energy and mass flows with little or no regard to the socio-economic, technical, emotive and political dimensions of sustainability [24]. It further revealed that individual country teams establish scoring weights subjectively when evaluating building products, which often pose problems when applied to other regions [25,26]. The analysis of the study however, showed little evidence to justify the assumption that there are tools of demonstrable reliability for designers to assess the sustainability and suitability of such materials or products or their applicability and utility for their potential use in the design of low-cost green housing projects. Hence, a more reliable method is needed to aid design and building professionals in the selection of such building materials and components for low-cost green residential housing projects.

Consequently, to promote more informed decision-making in the selection of low-cost green materials both individually and as assembled building components, a Decision Support System (DSS) is presented in this paper as an aid to design and building professionals. The objective of this study is to support decision-makers in selecting low-cost green building products that are environmentally, socio-culturally, technically and economically balanced through a proposed conceptual system. The model is to facilitate the integration of more sustainable materials into future designs by helping designers quantify how they compare to materials already permitted under existing codes, using the concept of the Analytical Hierarchy Process (AHP). The AHP approach is designed to be practical, as it combines environmental, technical, socio-cultural and economic performance into a single performance value that is easily interpreted.

In the following sections, the reviews of existing technological approaches are summarised and the main findings and themes to emerge from the literature review and the fieldwork seminars and interviews are reported. Then a step-by-step methodology is presented to illustrate the different stages of the DSS model development. Finally, the application of the prototype DSS for selecting appropriate floor material for a residential project in the London Borough of Sutton is demonstrated. The final section concludes the study and suggests areas for further research.

2. Technology in Material Selection: Review

For the past ten years, proven and commercialized technologies have been developed to promote environmental awareness amongst built environment professionals [18,22,23]. Empirical research validates that various studies on building material selection support systems have developed in size and specification within the last ten years [24-26]. Castro-Lacouture *et al.* [26] note that the application of green building support/assessment tools has been widely accepted as an effective and useful way of promoting green housing construction in the housing construction industry. Keysar and Pearce [27] and Bayer *et al.* [28] however, argue that the contexts in which building environmental assessment methods now operate, and the roles that they are increasingly playing, are qualitatively different than earlier expectations. They note that material assessment tools are now classified based on the type of analysis they perform, such as product, assembly, or whole building analysis, or classified as region-specific tools, either considered based on the life-cycle phases they cover, or on the required skills necessary to operate the tool.

While there is clearly an urgent need for new technologies to optimise the use of low-cost green building materials, it is also true that there are many technologies or systems already in use [25-28]. The first real attempt to establish a comprehensive means of simultaneously assessing a broad range of environmental considerations in building materials was the Building Research Establishment Environmental Assessment Method (BREEAM) [28]. The BREEAM tool assesses the environmental impacts of over 150 various materials and components most commonly used in home construction. The tool takes environmental issues into account, then adds measurements and user-defined weighting to arrive at environmental impacts, measured as "Eco-points" for each building material being assessed. Twelve different environmental impacts are individually scored, together with an overall summary rating, which enables users to select materials and components according to overall environmental performance over the life of the building. This scientifically accepted program however, focuses only on the environmental performance of products rather than environmental, social and financial considerations going hand in hand as parts of the material evaluation and selection process.

With emphasis on the Leadership in Energy and Environmental Design tool (LEED), Keysar and Pearce [27] conducted a detailed evaluative study comparing the effectiveness of five different relative importance indices for selecting appropriate material selection tools such as: relative advantage; compatibility; complexity; trialability; and observability, with the goal of improving the sustainability of materials for capital projects. Here, materials such as; regionally manufactured materials, materials with recycled content, rapidly renewable materials, salvaged materials, and sustainably forested wood products are selected based on credit scores. Analyses of their study however, revealed that the LEED model for example specifically requires an energy model, a task often handled by a specialist within a design firm or outsourced to a third party specializing in energy modeling.

Due to the inflexibility inherent in the application of first generation tools, and since they tend to require greater technical expertise to implement, many different tools of the second generation group have also been launched to address these limitations. Among this category is the ATHENA estimator. This has been one of the most popularly used material data-analytic models that analyses over 1200 building material and assembly combinations [28]. It allows the users to look at the life cycle environmental effects of a complete structure or of individual assemblies and to experiment with alternative designs and different material mixes to arrive at the best scenario. Bayer *et al.* [28] noted that the major drawbacks to this tool are the fixed assembly dimensions, software cost, the cost and required skills to use it, the limited options of designing high-performance assemblies, and the overall incomplete assessment of whole buildings environmental impacts [28-30].

With the identified setback associated with ATHENA

estimator, The National Institute Standards and Technology (NIST) developed the Building for Environmental and Economic Sustainability (BEES®) 4.0. This model provides a cradle-to-grave product-to-product comparison of over 230 building products based on manufacturer and supply company information [28-30]. The impact categories are weighed, normalized, and merged into a final environmental performance score, to generate a single measure of desirability for product alternatives by combining qualitative and quantitative data. The BEES 4.0 model is however, not capable of providing data for a full LCA of a complete building product, as it only produces data for a limited amount of building materials and evaluative factors [28-31]. These single-attribute claims ignore the possibility that other life-cycle stages or environmental impacts can yield offsetting impacts. Other limitations include; limited product options, limited use for local/regional impact materials and de-valuating weighing process [17].

Trusty [18-21] argued that these sets of first and second-generation tools less often consider any of the Multi-Criteria Decision Methods available to solve MCDM problems, adding that some systems do not even consider Life Cycle Cost (LCC) and other performance criteria simultaneously or completely. Moreover, he claimed that the existing performance requirements/criteria approach used in such tools tend to rely on immeasurable characteristics in demonstrating the extent of sustainability in a product, which makes them over-burdensome to implement and communicate.

Since the highlighted material assessment tools were developed primarily to be used in different countries, and the data sources used by each tool differed, further efforts have been undertaken to develop knowledge-based or expert DSS for assistance in material selection. For instance, Rahman et al. [32,33] developed an integrated knowledge-based cost model for optimizing the selection of materials and technology for residential housing design using Technique of ranking Preferences by Similarity to the Ideal Solution (TOPSIS). The system is developed to assist architects, design teams, quantity surveyors and self house builders to make decisions for the design from early stage to detailed design stage by ranking the performance and cost criteria of technologies and materials. Loh et al. [34] however, criticised the tool for providing partial assistance in the material selection process of the whole building design as it only considers the cost of roofing materials. They argue that material selection process depends on a number of other factors such as the location, zoning and environmental regulations, demographic characteristics, etc. that are not considered in their system. They note that the TOPSIS approach adopted does not only lack the ability to eliminate bias in the selection process but also unable to allow fairer

trade-off process.

Loh et al. [34] emphasise that strategic selection of sustainable materials and building design prior to the building construction is crucial to increasing building life cycle energy performance. They argue that stakeholders involved in the early design process often have conflicting priorities for both building design and construction materials. They developed an environmentally focused decision support system in the form of an Environmental Assessment Trade-off Tool (EATT), which supports the development of the ideal building design and materials combination that meets stakeholders' requirements. It is designed to assist users select the most appropriate material among a set of candidate materials based on the analytical hierarchy process (AHP) concept of decision-making, since AHP technique has the robust ability to handle the complexities of real world problems, and to deal formally with judgment error, which is distinctive of the AHP method. The system rank orders a set of preselected, technically feasible materials using different decision factors with and without tangible values, such as a clients favour over a particular building design, publicity potential of the building design, life cycle cost, capital cost and energy performance of different materials and building layouts. Zhou et al. [12] argued that the approach adopted by Loh et al. [34] lacked in robustness as it does not take into account the full-life cycle impacts of newly-accepted building products, and did not specify the sort of materials under studied.

Zhou et al. [12] developed a decision support multi-objective optimization model for sustainable material selection. The material selection tools and material data sheets provide extensive information that includes factors such as cost, mechanical properties, process performance and environmental impact throughout the life cycle based on expert knowledge. Wastiels et al. [16], confirmed that the tool, however, lack the considerations or descriptions to evaluate the intangible aspects of building materials, which are important to architects. They also criticised the selection methodology for being highly restrictive to a limited range of factors and incompatible with other stakeholders.

Ashby and Johnson (2002) introduce "aesthetic attributes" in the material properties list for product designers when describing material aspects such as the transparency, warmth, or softness. Within the discipline of architecture, however, the intangible qualities of materials are not described and mapped within the current design models. No selection framework was provided to support the implementation of a system.

Wastiels et al. [16], proposes a qualitative and quantitative framework to support informed decisions based on physical aspects' and "sensorial aspects" of building materials, but without the tools integration and computerisation as done by Zhou *et al.* [12]. In the presented framework, no pronouncement is made upon how sustainable considerations from these different categories could influence each other, and what MCDM approach could possibly be used if developed.

A similar study by Ding [35], developed a comprehensive assessment decision support system that measures the environmental characteristics of a building product using a common and verifiable set of criteria and targets for building owners and designers to achieve higher environmental standards. Upon analysis it was found that the assessment for her study focused heavily on environmental issues rather than the broader social, cultural, technical and economic aspects of sustainable green construction.

Keysar & Pearce [27] cited extensive research literature describing how material selection tools facilitate the innovation diffusion process and radical decision-making transformation. They however, note that most of the examined models make choices that result in "fabricated assemblies of standardized performance attributes", implying that they do not choose for materials but rather for 'material systems'.

Hopfe *et al.* [36] conducted a study that assessed the features and capabilities of six software tools to screen the limits and opportunities for using BPS tools during early design phases. The tools classification was based on six criteria namely the capabilities, geometric modeling, defaulting, calculation process, limitation and optimization. However, the authors did not report what methodology was used to compile these criteria.

A cost modeling system for roofing material selection was further proposed in Perera and Fernando [37]. Several factors were identified and considered in the selection process. Results demonstrated large inconsistency in the evaluation process. No particular reference was made to the selection methodology.

Other influencing reviews within the scope of this study include Mohamed and Celik [38] who proposed a computerised framework that is responsible for materials selection and cost estimating for residential buildings where users are able to choose their preferred one from list of materials without evaluation and synthesis of multiple design criteria and client requirements. No mention was made about the MCDM technique used for evaluating the list of materials selected and their respective quantities.

Mahmoud *et al.* [39] suggested a method for the selection of finishing materials that covered floors, walls and ceilings and integrates cost analysis at the appropriate decision points, but without the selection information requirements or methodology as proposed in this study.

Lam *et al.* [40] carried out a survey on the usage of performance-based building simulation tools. His study examined the relative impacts and limitations of knowledge-base tools in decision-making. Murray argues that while there is a natural tendency for design and building professionals to focus on the scientific and technological aspects of green and sustainable construction, their approach does not necessarily maximise the positive contributions professionals have to offer if tools are designed to replace professional judgment in the choice of materials. Murray suggests that this is because tools cannot address the intrinsic motivations people need if they are to embrace the positive changes sustainability requires. He continues that limiting the assembly of buildings to the specification of systems would impede the discovery of design opportunities inherent in materials themselves. Similar patterns of consistency, and lack thereof, have also been obtained [for detailed reviews see 17,24-27,31].

By highlighting the different green building material assessment tools, it can be deduced that existing tools are dispersed and based on individual initiatives without a unified consensus based framework [41,42]. It is apparent that each tool has its own unique application. While each tool could be called an LCA tool, there was little consistency in the methodologies used from one tool to another. In addition, while one tool considered the building as a system, other tools considered primarily the product's individual attributes rather than how that specific product performed within the building system [42]. A key question therefore, is whether current assessment methods that were conceived and created to specifically evaluate the environmental merits of conventional building materials can be easily transformed to account for a qualitatively different set of materials.

Giorgetti & Lovell [43] for instance have reported the sub-optimal performance of existing tools. They argued that the subjective values and priorities of the authors of the assessment scheme largely dictate the technical characteristics of the systems, and currently represent the major focus of discussion. They suggest that it is necessary for potential users to analyse the local situation and identify the adaptability of using any tool before applying a universal green building assessment tool to a specific country and region. They warned that some existing tools such as BREEAM, LEED, and even current expert tools might potentially institutionalize a limited definition of environmentally responsible building practice at a time when exploration and innovation should be encouraged in another region.

However, in all the reviewed studies, no efforts to develop a DSS that associates with the corresponding attributes and performance characteristics of low-cost green building materials and components, starting from the broad list of available options in the database to the final selection of the most appropriate material, were found in the existing literature [43,44].

The findings of the review have shown that each of the indices applied in developed regions to deal with issues associated with the impacts and performance of low-cost green building materials in other regions have proven unsatisfactory [44,45]. This finding is premised on the fact that most existing material selection systems have been designed by countries with more developed economies such as the UK, where the scale of social issues and lack of access to resources is simply not as critical as observed in the developing nations [45,46]. The setbacks that associates with the tools reviewed in this research thus, highlights the opportunity for developing a Material Selection Decision Support System (MSDSS), to better address the specific needs and attributes specific to the use of low-cost green materials for tool adopters new to green housing.

The following section briefly highlights the aim and objectives of the study. It extensively describes specific methods adopted for each task in Section 3.1.

3. Research Methodology

In order to identify the key selection factors or variables that formed the basis for the development of the proto-type multi-criteria decision support system (DSS), suitable clusters of research approaches were considered in the research exercise, some of which include: exploratory literature reviews, networking with domain experts and practitioners, series of questionnaire surveys and knowledge-mining interviews [47]. **Table 1** provides an overview of the research aim, objectives and the methodology undertaken in four major stages.

3.1. Research Design

To provide a clear theoretical framework for the relatively new area of study, and develop preliminary ideas on issues specific to the research theme within the context of decision-making associated with the impacts of low-cost green building materials and components in housing construction, this study reviewed relevant literature through synthesis and analysis of recently published data, using a range of information collection tools such as; books, and peer-reviewed journals from libraries and internet-based sources. Recognising the limitations of the literature review in terms of examining current research thinking in respect of decision support systems for the selection of low cost green building materials and components, a preliminary research study was undertaken to check and validate prior assumptions in the background and review sections.

In order to build upon knowledge gained from the literature review, and recognising the limitations of the preliminary research survey in terms of examining current research thinking in respect of decision support sys-

tems for low cost green building materials and components, a mixed method was adopted for this study. This was followed by in-person interviews to further clarify and elaborate on less detailed and pertinent issues associated with the use low-cost green building materials. The in-depth interviews consisted of 10 participants, who involved a sample of practicing architects, engineers, material specifiers, and a host of building professionals-who influence material choice decisions in the UK housing construction industry. This approach was used to examine the potentials of the proposed MSDSS, (being a tool for the assessment and evaluation of low-cost green materials). It further investigated the effectiveness of design and decision support tools, as well as identified requirements of Life Cycle Assessment (LCA) tools for design decisions at the various stages of the design process.

Consequently, a quantitative questionnaire was developed as the result of the analysis of the results from the interviews. In order to elicit the "most important" factors, a questionnaire survey was conducted among the executives of some selected builder/developer firms. They were asked to rank order from a list of factors (compiled from existing literature on the topic and after initial consultation with some of the executives) based on their judgment and experience. The executives were also asked to indicate desired features they would like to have in a DSS for low-cost green material selection. Since the respondents were widely dispersed, and because it was anticipated that building professionals would be more likely to reply and cooperate with a less time-consuming research method, giving the constraints of time, wider coverage, and budget, it was therefore, decided that a questionnaire sent and returned by email would be the most convenient way of collecting the required data. The inclusion of qualitative open-ended questions provided respondents a chance to express their views more freely.

The target groups of respondents were also taken from a database or directory of building professionals provided by the UK, China, Canada, South Africa, Brazil and US Green Building Councils (GBCs). The selection approach followed the random sampling technique to avoid bias and uneven sample sizes amongst different professional groups, and ensure uniformity, consistency and quality of data. To facilitate the response rate, snowball sampling was also adopted, where the approached respondents were asked to distribute the questionnaire to their colleagues and partners within the field [47].

The selection of South Africa and Brazil for the analysis was due largely to their great similarities in social, economic, and geopolitical terms, and likewise their developed counterparts. In a similar vein, the choice of building experts within the selected countries was as a result of their expertise and advancement in the use and

Table 1. Basic summary of the research methods.

Stage	Objectives	Tasks	Method
AIM (spanning)	To develop a decision support system (DSS) that will provide designers with useful and explicit information associated with low-cost green building materials and components, to aid informed decision-making in their choice of materials for low-cost green residential housing projects.		
1: REVIEW	1.Examine current views on themes related to decision-making associated with the use of low cost green materials in the housing industry, to identify new ideas & issues arising from the study	**Step 1.** Reviewed relevant literature through synthesis and analysis of recently published data, using a range of information collection tools such as; books, peer-reviewed journals, and articles from libraries and internet base sources	AA,
	2. Review various DSSs currently used at national and international levels for the selection of materials to identify knowledge deficits and the potential benefits associated with their use	**Step 2.** Carried out a preliminary research study with leading researchers who influence the selection of building materials in the field of housing construction	AA, QS, INT
2: DATA COLLECTION &SYNTHESIS	3. Conduct surveys and interviews with building professionals, to identify the potential factors or variables that influence the informed selection of low cost green building materials and components	**Step 3.** Conducted a pilot study, by deploying a test-questionnaire to a small sample of researchers who possess relevant knowledge on issues specific to the use of low cost green materials using the email addresses taken from the databases of recognised building construction companies and research institutions	AA, QS, INT
		Step 4. Conducted the main survey, by administering the revised questionnaire through email contacts taken from databases of interested registered building professional groups, who influence the selection of construction materials from throughout the construction value chain	
		Step 5. Conducted in-person interviews with interested building professionals who influence material choice decision in housing construction using audio recording system to avoid re-contacting the respondents or falsification of information	
		Step 6. Carried out inspection on available expert systems most commonly used in building firms in the UK, USA, China etc. by interviewing experts, with years of experience in the industry, who have implemented or used such systems and directly observing how they function when in operation	
3: DATA ANALYSIS	4. Evaluate and establish the weighted importance of the key factors or variables that will help to determine the relative impacts of the different choices of building materials and components	**Step 7.** Analysed the information and report gathered from the survey exercise(s) using a suite of statistical analytical programs, and various quantitative data analytical techniques	AA, QS M
	5. Develop a system to integrate the necessary information appropriate to the informed selection of low-cost green building materials & components	**Step 8.** Assembled the key components by synthesising the relevant databases to be incorporated in developing the proposed DSS model.	AA, QS, M
		Step 9. Developed the main structure workflow of the proposed system by creating links among the various databases,	
		Step 10. Inputted relevant data to test the internal links to know what needed to be measured within the system, and checking the output of the results against easily calculated values	M
4: DEVELOPMENT	6. Test the functionality of the proposed approach; and validate the effectiveness by applying it to a building material selection problems using a series of case study residential building projects in the UK	**Step 11.** Conducted experts survey by deploying a sample of the prototype system via email of those who participated in the main survey, using feedback questionnaires as a quicker and cost effective means of assessing respondents' judgments about the system	QS
		Step 12. Made necessary changes based on the feedback from the survey	M
		Step 13. Validate the modified prototype system using a series of completed building projects in the UK, by comparing the outputs from the algorithms to monitored data from the completed building	M, CS

KEYS: AA (Archival analysis); INT (Interview); CS (Case study); QS (Questionnaire Survey); M (Modeling).

development of green building tools (as they have had the most uptakes in both geographical regions and being part of an emerging market).

To receive a reasonably sized sample, 500 surveys were sent out by email, over a two-month period of March and April 2012. Using a progressive approach of data collection, a total of 250 respondents returned the completed survey, representing a response rate of 50%. The response rate was accepted as the normal ranges between 20% - 30% were found in most of the construction industry related research [33,34]. Prior to distribution, the questionnaire was pre-tested for comprehensibility by consulting five academics at two universities [47]. A number of changes were suggested and implemented.

Respondents were also invited to post their ideas about current limitations or improvements that should be avoided or integrated in the development of the proposed MSDSS model at the later part of the questionnaire. The questionnaire also examined the adequacy/inadequacy between traditional manual approach of material selection and computer-aided decision support tools. One of the group's participants commented that one of the hallmarks of good science is that a result can be tested independently and proven to be right or wrong in the latter method. The analysis of the questionnaire survey and interviews provided a list of "key" decision-related factors having significant impacts on the process of material selection for residential development as shown in Section 4.1.1.

3.2. Research Findings

The results of the study however, revealed the following.
- Many existing decision support systems in the developed countries do not have the appropriate performance threshold for addressing the most relevant issues specific to less developed countries;
- Current DSS models are unable to relate to matters associated with the informed selection of materials that are commonly used for housing projects in countries with rather less-mature markets;
- The lack of informed knowledge by building professionals in terms of the principles, characteristics, and best practices relevant to the use of low-cost green materials at the design stage, has been identified as a common constraint peculiar to their wider-scale use in the housing industry;
- The majority of building professionals still regard cost and environmental factors as conventional project priorities when selecting building materials or components, but rarely consider the implications of social, political, technical, sensorial, legal and cultural factors in their choice of materials; and finally,
- The majority of low-cost green building materials are

yet to be certified under the building regulations, standard specifications and codes of practice; and most importantly,
- There are no demonstrable and compelling evidence of technical research on a holistic approach used by design professionals for the evaluation and selection of low cost green building materials and components at the design stage.

The results of the study thus, provided the platform that suggested the need for a system that could aid informed decision-making to improve understanding, and enhance the effectiveness of actions to implement and promote the wider-scale use of low-cost green building materials and components at the core of the construction business process. In light of their feedback and useful suggestions from building experts who partook in the study, the following portions of the DSS model were either readjusted or improved.
- Easy searchable material selection inputs database;
- Ability to add/remove material selection features with ease;
- Ability to make custom reports;
- Ability to easily navigate all components with ease;
- Comprehensive "HELP or USER INSTRUCTIONS" menu explaining what the tool is doing;
- Being able to understand the material selection process through the lens of non experts;
- Ability to perform trade-off analysis to compare different material options;
- Clarity on the algorithms used to perform the simulations; and Real-time results;
- Data input forms to ensure easy and consistent data input; and,
- Having a huge amount of customizability in terms of output.

After the improvement, the system was shown to the same participants, and minor adjustments were made on the basis of second feedback. In the following sections the proposed MSDSS selection methodology is discussed, and a conceptual framework for the decision support system based on the methodology is presented. Subsequently, the MSDSS model is applied to a hypothetical but realistic material selection problem to rank order the candidate materials for selecting the most appropriate one.

4. System Development

For this research, AHP was selected for its simplicity and due to the fact that it can be easily implemented using any spreadsheet software application such as the MS Excel, as it possesses a powerful macro language that is essential since a menu driven interface had to be developed. Since the intention of the research was not to develop a commercial software product, Macro-in-Excel

VBA (MEVBA) was utilized for the following reasons:

- Macro-in-Excel VBA (MEVBA) has the capabilities to perform all necessary calculations and is common enough that most people are familiar with it;
- It has the ability to write scripts that could automatically convert material data from any graphic table format to an appropriate condensed data table (hidden from the user's view) to allow quick and reliable indexing of material data;
- The Macro-in-Excel VBA framework has the code that makes Windows forms work, so any language can use the built-in code in order to create and use standard Windows forms;
- Makes the application easier to maintain; With MEVBA, codes were easily built into the form or report's definition, since the DSS model contained a large number of macros that respond to events on forms and reports; which would have been difficult to maintain using any other application;
- With Macro-in-Excel VBA it was easy to step through a set of records one record at a time and perform an operation on each record;
- Macro-in-Excel VBA helped to supply a standard security mechanism, which was made available to all parts of the MSDSS data application model;
- Enables the developer to create his own functions: The MSDSS contains a series of mathematical model and computational algorithmic procedures that provided a basis for computing the green development index of material alternatives within an integrated decision-support framework or tool(s).
- Ability to mask error messages during the tests run;
- Enables the system to quickly analyze existing data to discover trends so that predictions and forecasts can be made with reasonable accuracy;
- Allows for extensions and expansions: since the components of the framework are modular, meaning that each may be developed independently, and data may be added as it is acquired to supplement the knowledge and databases, macro-in-excel was used to achieve that goal

4.1. MSDSS Database/Data Warehouse Design

The data warehouse design constitutes the major portion of the MSDSS development and hence will be explained in detail in this section. The data warehouse design essentially consists of four steps as follows:

Step 1: Identifying the key influential factors that will impact on the choice of materials;

Step 2: Designing the material selection methodology framework and identifying the objectives of each step;

Step 3: Designing the various components of the MSDSS model and defining their features and functions;

Step 4: Defining the workflow selection methodology and analytical procedure of the actual prototype MSDSS model

4.2. Identifying the Key Influential Factors

In order to identify the relative importance of the subcategorical factors or variables based on the survey data, ranking analysis was performed. Five important levels were transformed from Relative Index values: Highly Significant Level (H) ($0.8 \leq RI \leq 1$), High-Medium Level (H–M) ($0.6 \leq RI < 0.8$), Medium Level (M) ($0.4 \leq RI < 0.6$), Medium-Low Level (M–L) ($0.2 \leq RI < 0.4$), and Low Level (L) ($0 \leq RI < 0.2$).

From the results of the analysis, 40 factors were identified under the "Highly significant" level for evaluating low-cost green building materials with an RI value ranging from 0.952 to 0.806 and a total of 15 factors, were recorded to have "High-Medium" importance levels with an RI value ranging from 0.795 to 0.652. The analysis of the main survey identified a total of 55 key influential factors out of 60 initial factors as important components of the material selection process.

"Life Expectancy" was ranked as the first priority in the technical category with an RI value of 0.952, and it was also the highest among all factors and was highlighted at "High" importance level. "Resistance to fire" was also rated high in importance among the selection factors. "Maintenance Cost" was ranked third in importance. It was clear from this research that there is a perception of ambiguity surrounding the long-term maintenance of low-cost green building materials. This is not entirely any surprise given that maintenance free buildings are increasingly sought after by clients, anxious to minimise the running costs associated with buildings. "Life-cycle cost" has been, and will continue to be, major concerns for building designers, as well as important traditional performance measure.

Among the top 20 ranking factors, it was observed that only one factor from the environmental category out of the list was ranked high among the selection factors. This again suggests that environmental issues within the context of the developing countries are not strongly considered despite the high environmental awareness exhibited by design and building professionals in developed regions. This finding also corroborates the initial observations of various studies [14,15] repeatedly highlighted in the background and literature studies. They suggest that the problems within the developing regions are characterised by mainly social and economic issues, unlike the developed regions where the scale of social issues and lack of access to basic resources are simply not much of a problem as it is in the developing world.

From **Figure 1**, a total of 15 factors, consisting of 12 site factors, 1 socio-cultural factor, and 2 sensorial factors, were recorded to have "High-Medium" importance levels. Although these 15 variables were in the same importance level category, the "building orientation" factor within the "general/site category" (average RI = 0.652) was considered to be the least important variable compared to the factor "Glossiness" under the "sensorial category" (with an average RI = 0.774), and "material availability" still under the "general/site category" (with an average RI = 0.795). However, it should be noted that site factor accounted for 75% in the "High-Medium" importance level. The result is an example of evidence pointing to the trend that environmental and perhaps site issues are no longer considered as the most important factors for material selection in housing projects, especially within the context of the less developed regions.

Some factors in the three categories were ranked relatively higher in the "High-Medium" level. For example, "material availability (GS1)" was rated as first in the general/site subcategory, and ranked as thirty-fifth in the overall ranking with an RI value of 0.795. An interesting observation from the results is that none of the criteria fell under the medium and other lower importance level. This clearly shows how important the factors are to building designers in evaluating low-cost green building materials. All factors were rated with "High" or "High-Medium" importance levels. However factors such as Compatibility with other materials, Skills availability, and UV resistance fell within the medium-low level. The

findings of the analysis asserted that the criteria with medium or low RI does not mean they are not important for selecting materials, but rather created an opportunity to highlight the relative importance of the key criteria from their vantage points. The following shows a framework consisting of the key factors in their order of importance.

4.3. Designing the MSDSS Selection Methodology

The diagram shown in **Figure 2** demonstrates the conceptual framework of the selection methodology for the decision support system. **Table 2** describes a step-by-step procedure of the selection methodology for the material selection decision support system. Section 4.4 presents various components of the MSDSS schema or model.

4.4. Designing the Features of the MSDSS Model

The next stage of the model development was to design the various features of the databases containing the logic and showing relationships between the data organized in different modules. Each module contains the physical information and contents needed to aid in the material evaluation and selection process.

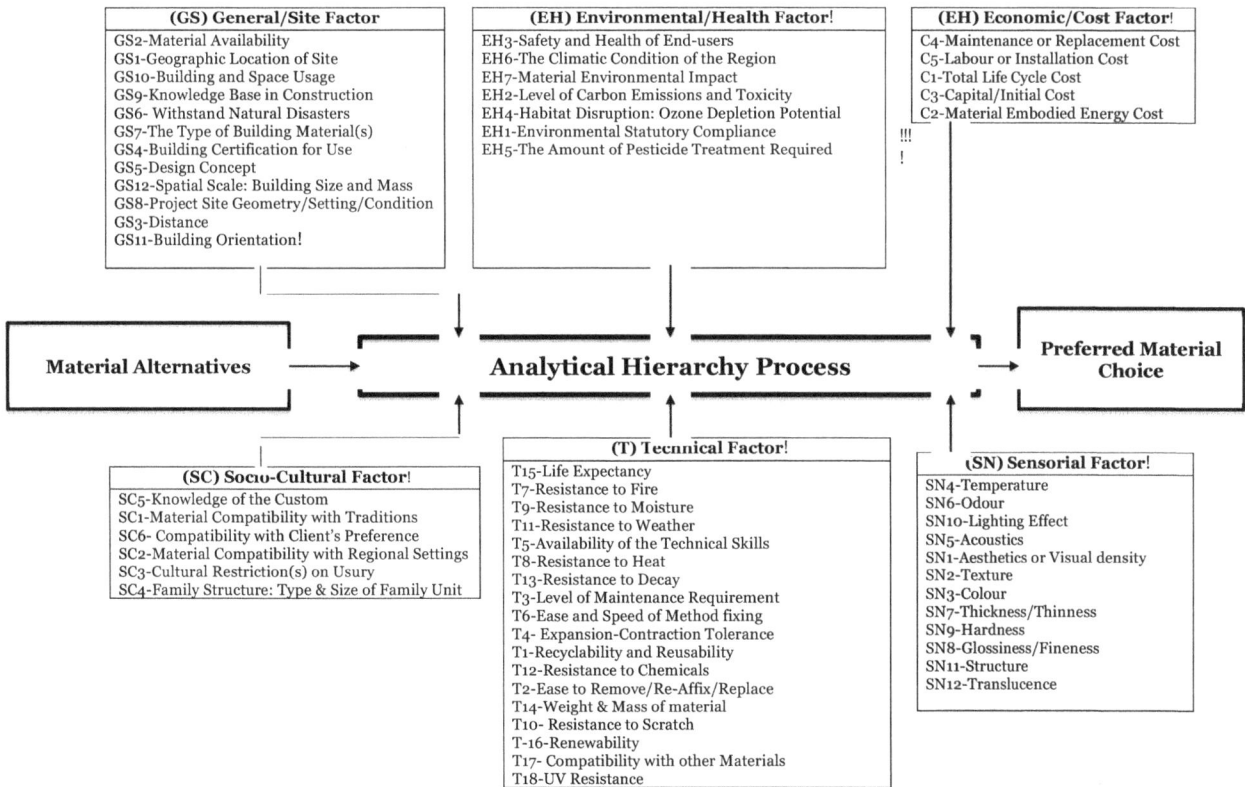

Figure 1. Ranked factors for measuring the impacts of low-cost green building materials.

Table 2. Description of the selection methodoly.

OBJECTIVE	TASK
1. Define or state overall objective/goal	The first step of the methodology is to define the main goal of the intended task.
2. Identify Set of all Possible Material Alternatives to be Assessed	After defining the main goal of the task, the next step is to generate the set of all possible alternatives that are available for selection based on the decision-making parameters. In the material selection process, this comprehensive set of alternatives includes all construction materials and components currently in the database, and the market in context.
3. Prune all infeasible alternatives from set	The third step is to reduce the complete set of alternatives by eliminating/pruning those alternatives, which are clearly infeasible for the intended application from the database consisting of all materials, based on classifications of materials according to the Construction Standards Institute (CSI) Divisions, and material heuristics. For example, if the element under consideration is a structural beam, materials such as roofing sheet and glass are automatically pruned from the set of possible alternatives under consideration, since none of these materials fall under the CSI structural divisions. This should result in a subset of alternatives, all of which would be feasible choices for the intended application. The "pruning" approach is used rather than allowing the user to select feasible materials from the whole set because users tend to overlook alternatives which might be unfamiliar to them but are nonetheless feasible.
4. Evaluate Remaining Alternatives	The fourth step in the methodology is to evaluate the feasible alternatives using the AHP model such that a ranking can be developed according to the relative importance of the material for the intended application.
• Weight Attributes (Decision Factors)	• First, the decision maker weights each factor or variable according to the relative importance that the decision factor or variable holds for the decision maker. It involves the decision-maker replacing probabilities with user weightings for each factor or variable to supplement, not replace, his judgment.
• Calculate Values for Attributes	• Second, values for each of the factors or variables are determined for each material with regard to the manufacturer's information & details of the material or component contained in the material database, and then, a normalized value between zero and one is calculated for each factor value. • After weights have been established and values calculated for each attribute against a set of materials or components, the weights and normalized values are multiplied and summed to create an index of preference for that alternative(s).
• Amalgamate Weighted Attributes • Develop Ranking	• Then, a list of alternatives ranked according to the relative importance of the factors or variables is then presented.
5. Review Ranking of Alternatives	When the indices of factors or variables have been calculated for all feasible alternatives, a ranking is developed sorting the alternatives according to each utility value based on the AHP model of decision-making. The alternative with the highest utility value is recommended from the ranked list of potential materials for each design/building element.
6. Select Alternative Based on Ranking	The decision maker may then either elect/decide to select the highest ranked alternative, or choose another alternative from the set based on his professional judgment.
7. Proceed to Next Design Elements	The decision maker satisfied with the selection process, then proceeds to the next design/building element.

The conceptual model/framework of the prototype MSDSS tool consists of a number of interconnected modules/features. A logical model illustrating the developed DSS for material selection is shown in **Figure 3**. **Table 3** describes the functions of each component of the MSDSS model.

4.5. How the System Works

The following steps explain how the prototype MSDSS model works during the material evaluation process.

Step 1: The load manager provides the user with a list of design elements from the "Design Elements" module, and then prompts the user to select the design element of his/her choice in accordance with the terms and specifications of the Construction Standards Institute (CSI) Divisions;

Step 2: The User then selects the particular design element needed for the intended task from a list of design elements (as broken down by the Construction Standard Institute Division);

Step 3: User then enters values for the relevant parameters to answer prompts about areas and dimensions

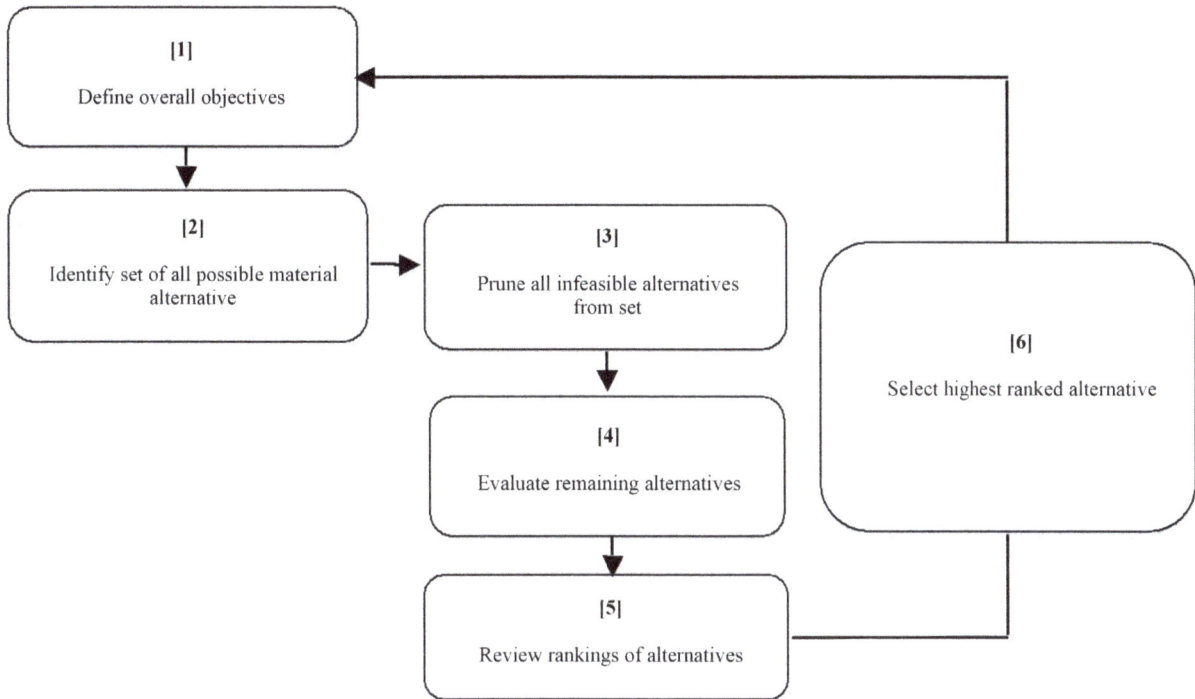

Figure 2. Selection methodology for the MSDSS model.

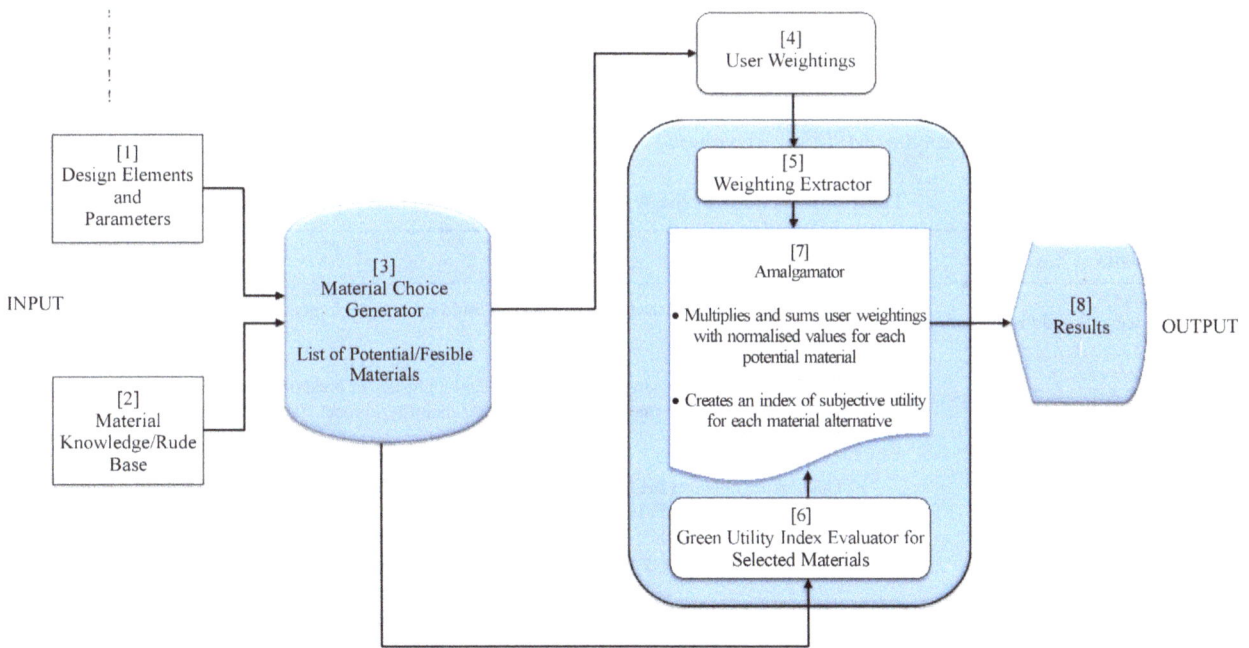

Figure 3. Conceptual framework of the MSDSS model.

of the selected design element, and then sets the threshold values in the material knowledge base

Step 4: The system validates the design parameters and threshold details entered by the user, and then generates the set of all feasible material alternatives that are available for selection, (which includes all categories of construction materials contained in the materials database);

Step 5: After a set of feasible material alternatives has been generated for the "particular design element", the system through the "Weighting Score Extractor Module" prompts the user to obtain weightings for the desired parent and sub-factors according to the relative importance that each factor or variable holds over another based on the decision maker's preference of value;

Step 6: After weights have been established and values calculated for each factor for a particular material, the weights and normalized values are multiplied and summed to create an index of subjective utility for each alternative;

Step 7: The alternative with the highest utility value is recommended by the system;

Step 8: The user reviews the system's recommended choice for each element in the "Result" module, and then either selects the highest ranked alternative, or chooses another alternative from the set based on professional judgment and/or the system's recommendation.

Step 9: The user may choose to generate a printout report or graphical representation of the list of selected materials and green utility indices if desired.

Step 10: The selection process then proceeds to the next design element.

Figure 4 presents a graphical representation of the system workflow.

An illustrative example of the AHP concept is displayed and explained in following section to demonstrate the selection process by applying the prototype MSDSS model to a hypothetical case study design project.

5. Application

The following example illustrates the selection process of floor covering products. It selects the best one among three alternatives. The prototype MSDSS, developed using the AHP technique, was used to select the most appropriate residential building floor material for housing development in the city of London, located in the Sutton County of London. The results demonstrate the capabilities of the MSDSS system in a real-life but hypothetical application scenario. In the following section this process of application is described and discussed.

5.1. A Hypothetical Study Case

The next stage of the model development was to design the various features of the databases containing the logic and showing relationships between the data organized in different modules. Each module contains the physical information and contents needed to aid in the material evaluation and selection process. **Table 4** summarizes the details for the three options of flooring materials for the proposed residential low-cost green housing project. The description of the three options in **Table 4** was based on the standard practices and construction details commonly used in the housing construction industry.

These three (3) floor materials described above will be analysed amongst a host of other material alternatives for the selection of a more sustainable option. In other words, this section will analyse the problem using the MSDSS model, which relies on the use of the AHP mathematical multi-criteria decision-making technique, to identify and decide which material is the most sustainable and suitable flooring material in this case.

To achieve this goal, the MSDSS model was sent to 10

Table 3. Functions of the features of the MSDSS model.

MSDSS Features	Functions
1. Design Elements and Parameters	This feature provides users with a range of building design elements and their respective parameters
2. Material Rule Base	This feature articulates the listing of individual materials in prescribed sequences, gradually eliminating candidate materials based on their inability to meet stated material selection heuristics/rules.
3. Material Choice Generator	This feature contains the material/component database, which generates the set of all possible material alternatives that are available for selection.
4. User's Weightings	Sets preferred weighting value for all attributes to compare with.
5. Weighting Extractor	This feature queries the user to obtain weightings for the factors, based on the user's preference of value on a scale of 1 - 9.
6. Material Index Evaluator	The material index evaluator calculates values of the selected factors or variables for each feasible material choice.
7. Amalgamator	Here the user's weightings are amalgamated (*i.e.* multiplied and summed) with the factor values or weightings for each potential material, resulting in a relative ranking of the feasible materials for each element.
8. Results	- This component provides the ability to view the processed data, and to generate reports. It allows the MSDSS model User Interface to communicate with the user; and also connects all the reports and queries that are generated in the Monitoring databases to the corresponding project files.

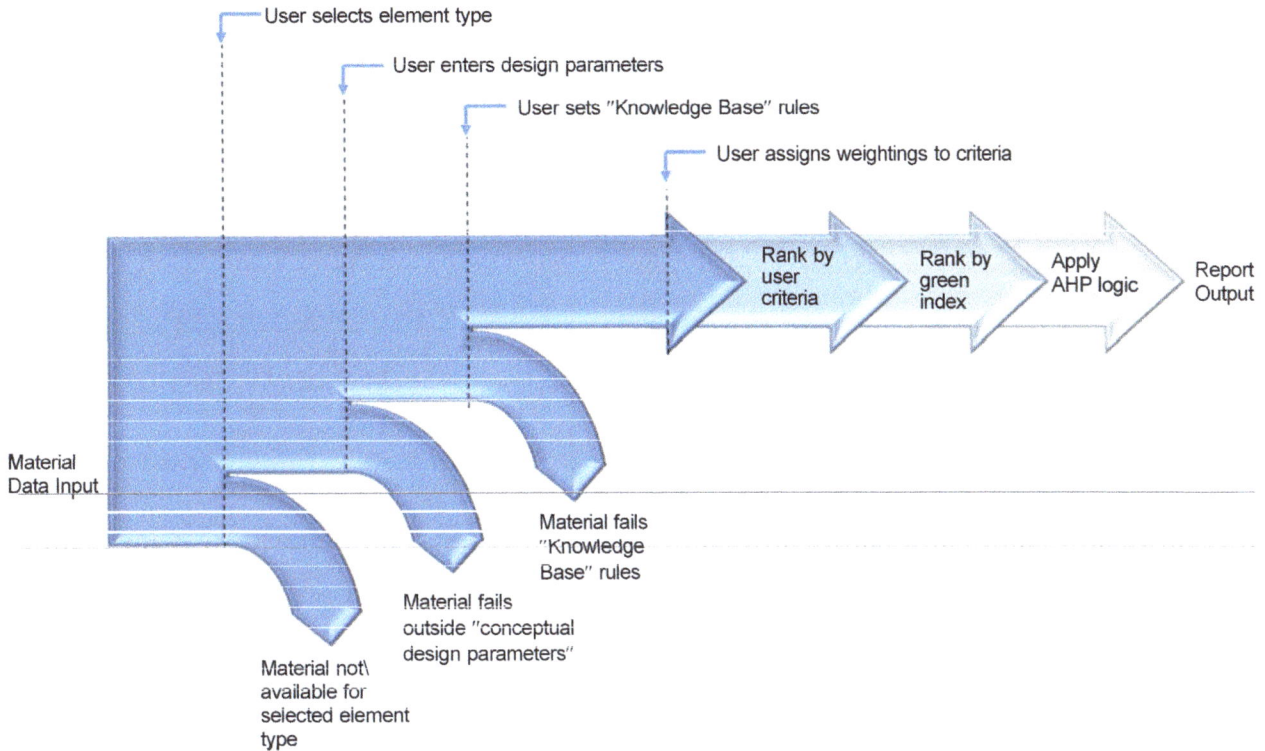

Figure 4. Workflow of the MSDSS model.

Table 4. Summary of the flooring options.

Description	Material A	Material B	Material C
Design Element Type	Paneled Flooring	Laminated Flooring	Concrete Flooring
Building Type	Residential	Residential	Residential
Material Type	Bamboo XL laminated Split Paneled Flooring	Reclaimed/Recycled Laminated Wood Flooring and Paneling	Fly Ash Cement concrete Floor Slab
Size of Materials	230 mm × 150 mm	50 mm × 6000 mm	900 mm × 900 mm

expert evaluators who had the following qualities:

- Considerable amount of knowledge in material analysis based on the AHP concept;
- Used a wide range of green building assessment tools for material selection; and
- Taken part in the previous survey.

The aim of this exercise was to compare their view of the prototype MSDSS model with existing models in terms of their usability, flexibility, and interoperability attributes using the concept of the Analytical Hierarchy Process (AHP).

5.2. Rationale for Adopting the AHP Concept

The study adopted the use of the AHP technique to investigate the interrelationships amongst various criteria and low-cost green material alternatives due to the following reasons:

- AHP is a method that is conceptually easy to use, and decisionally robust to handle the complexities of real

world problems;

- It does not require the very strong assumption that the stakeholders make absolutely no errors in providing preference information;
- It has the ability to deal formally with judgment error, which is distinctive of the AHP method;
- The AHP method provides the objective mathematics to process the unavoidably subjective preference inherent in real- world evaluations;
- Possesses an inherent capability to handle qualitative and quantitative criteria important for sustainable material selection; and finally,
- Can enable all members of the evaluation team to visualize the problem systematically in terms of parent criteria and sub-criteria.

Figure 5 shows the flowchart of the material selection computational analysis technique based on the concept of the Analytical Hierarchy Process model. The following sections present details of the evaluation exercise.

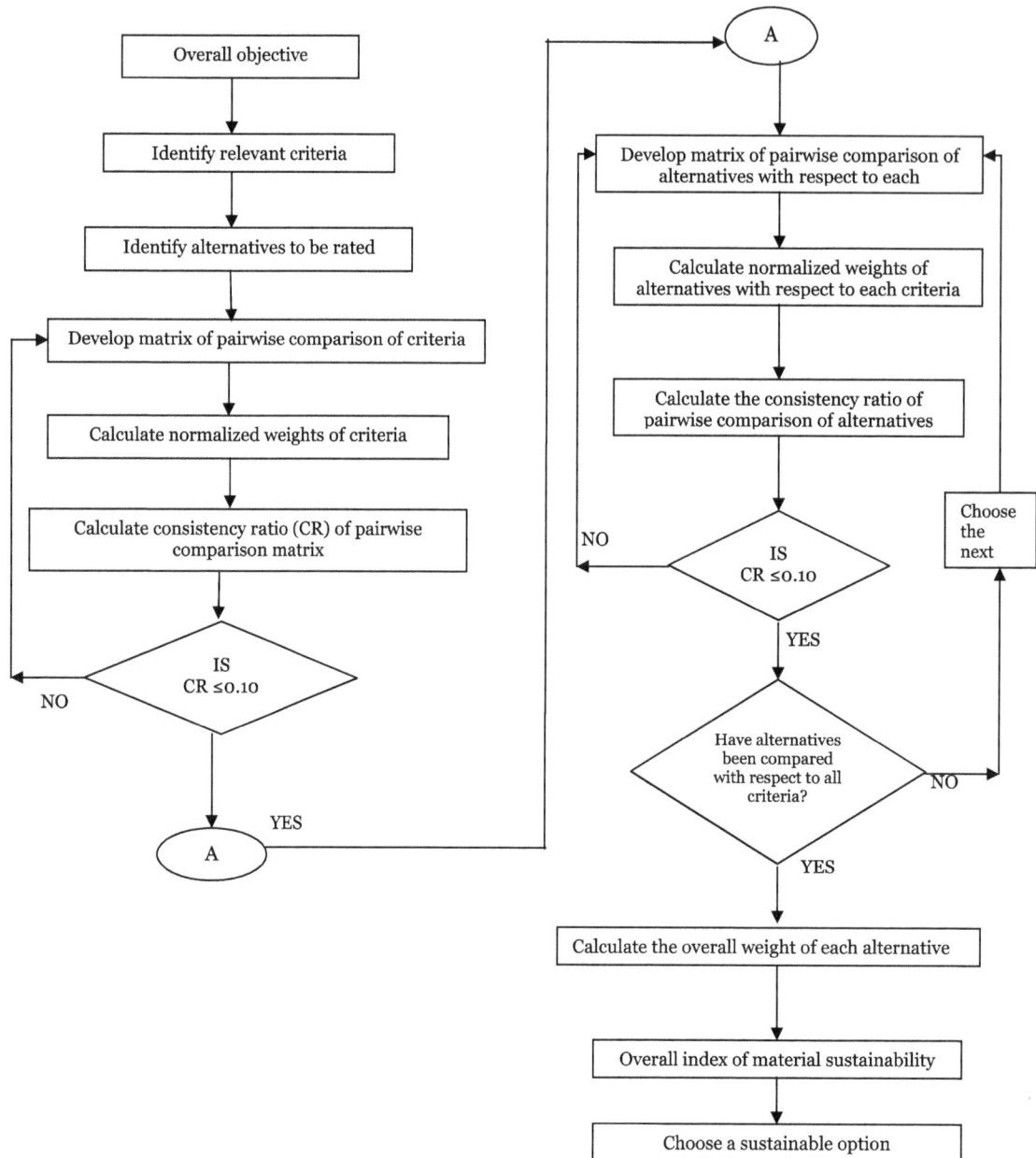

Figure 5. Flowchart of the AHP concept.

5.3. Applying the AHP Model to the Problem

According to Reza *et al.* [48], AHP is a subjective MCDM method that does not necessarily involve or rely on a large sample for its analysis. To better illustrate the procedure of the AHP technique of decision-making, with reference to the case presented in Section 5.1, a complete example of applying AHP to the problem of material selection is provided here based on evaluators' results. Twenty (20) respondents representing various fields of the housing construction industry, and who had fore knowledge of the AHP procedure were selected to participate in the AHP survey.

By evaluating the consistency level of the collected questionnaires, 5 questionnaires out of the 10 received had acceptable consistency and were entered into the system. In order, to avoid arbitrary and inconsistent answers in the data, the mean values of five (5) out of the ten (10) respondents were used to fill out the pair-wise comparison matrices for the parent and sub-factors.

The package included the model, evaluation questionnaire and a cover letter stating the purpose of the research, the validation process and what was expected of them. To conduct the exercise, the study adopted Chua's *et al.* [49] approach based on a number of suggestions as follows:

- A document that reminded and explained the overall aim and objectives of the study to the respondents, followed by a step-by-step demonstration of its operation;
- A demo illustrating a practical exercise. This allowed the evaluators the experience of using the system ensued. During the practical assessment session of the demo, evaluators were able to see the controls and get a general overview of the MSDSS interface;
- An illustrative example of the objective and methodology of the AHP technique based on the instructions in the demo, to guide and illustrate to every respondent on how to browse and conduct analysis;
- After the introduction, a feedback questionnaire was forwarded to the evaluators;
- After each evaluation, each evaluator highlighted their experience(s) and provided feedback on the feel of the system, with special attention to the problems that they encountered during the evaluation process;
- Finally, a reflective or post-user questionnaire was completed to obtain feedback;
- Evaluators were asked to answer each statement or question relating to the model in the questionnaire based on their personal view(s);
- They were also asked to assess the importance of the system based on their perception. Evaluators were also asked to add general comments on the system, and provide feedback on the applicability of the prototype system in assisting in specific material selection problems during their experience and other ways of improvement;
- Problems uncovered or areas that proved difficult to understand during the evaluation process were immediately modified so that it did not arise in subsequent sessions, as this procedure followed each evaluation;
- The respondents were instructed of the relevance of observing consistency in their answers whilst using the MSDSS model;
- The questions relating to different aspects were presented in different sections. This helped respondents to focus on one aspect at a time.

The following sections exemplify the process.

5.4. Decomposition of the Decision Problem

The evaluation exercise provided users with the opportunity to define the problem. **Figure 6** shows the exemplary hierarchy of the problem. The goal is placed at the top of the hierarchy. The hierarchy descends from the more general or parent factors in the second level to sub-factors in the third level to the alternatives at the bottom or fourth level as shown in **Figure 6**). To select a suitable choice among alternatives, the users were instructed to define the decision factors needed for the analysis. In other words, the users determined which alternative could be the best choice to meet the goal considering all the selected decision factors or criteria displayed in **Figure 6**.

The first step of the methodology (as illustrated in figure 2) was to define the main goal of the intended task, by identifying the design element needed for the analysis, and inputting the relevant dimensional scale for the suggested design element (see **Figure 7(a)**).

After defining the main goal of the task, the next step was to generate the set of all possible alternatives that were available for selection with reference to the decision-making parameters as shown in **Figure 7(b)**. At this stage the users are prompted or alerted by the MSDSS model to identify a set of feasible floor material alternatives based on a range of material selection heuristics/knowledge-based rules. The goal is to choose a suitable floor material among options for the project case described in Section 5.1.

5.5. Pair-Wise Comparison of Parent Factors

After selecting the design element, and identifying a set of feasible alternatives using the material selection heuristics/knowledge-based rules, the respondents were made to perform pair-wise comparisons following the demo instruction guide of the MSDSS model. This included the analysis of all the combinations of parent factors and sub-factors relationships. The sub-factors were compared according to their relative importance (based on the ratio scale proposed by Saaty [50-55], with respect to the parent element in the adjacent upper level. After performing all pair-wise comparisons by the decision-makers, the individual judgments were aggregated, basing its analysis on the geometric mean technique as Saaty suggested [52-55].

5.6. Pair-Wise Analysis of the Parent Factors

To avoid arbitrary and inconsistent answers in the data obtained from the 10 participants who consented to partaking in the study, the mean values of five (5) out of the ten (10) respondents were used to fill out the pair-wise comparison matrices for both the parent and sub-factors. The pair-wise comparison matrices obtained from 5 respondents were combined using the geometric mean approach at each hierarchy level to obtain the corresponding consensus pair-wise comparison matrices [54-56]. Using the verbal/ratio scale shown in **Figure 8**, respondents obtained weightings for each parent factor, based on the preference of value(s) on a scale of 1 - 9. The MSDSS model then automatically translated each of the matrixes into the corresponding largest eigenvalue problem and was solved to find the normalised and unique priority weights for each factor (as shown in **Figure 9**). Going by Saaty's [55] rule, the judgment of a respondent

Goal

SELECTING APPROPRIATE LOW-COST GREEN BUILDING MATERIAL

Main Factors

| GENERAL/SITE FACTOR | ECONOMIC FACTOR | ENVIRONMENTAL FACTOR | SOCIO-CULTURAL FACTOR | TECHNICAL FACTOR | SENSORIAL FACTOR |

Sub-Factors

GS1-Location	EH1- Env. Compliance	C1-Life-Ccycle Cost	SC1-Compatible (Tradition)	T1-Recyclability	SN1-Aesthetics
GS2- Availability	EH2-CO2 Emissions	C2-Embodied Energy Cost	SC2-Compatle (Region)	T2-Removability	SN2-Texture
GS3-Distance	EH3-Users' Safety	C3-Capital Cost	SC3-Control on Usury	T3-Maintenance	SN3-Colour
GS4-Building Certification	EH4-Ozone Depletion	C4-Labour Cost	SC4-Clients' Preference	T4-Stress Tolerance	SN4-Temperature
GS5-Designers' Experience	EH5-Pesticide Treatment	C5-Replacement Cost	SC5-Custom Knowledge	T5-Available Skills	SN5-Acoustics
GS5-Disaster Prone	EH6-Climate	C6-Maint.enace Cost		T6-Fixing Speed	SN6-Odour
GS6-Site Geometry	EH7-Env-Toxicity			T7-Fire Resistance	SN7-Thick/Thin
GS8-Spatial Structure	EH8-Fossil Depletion			T8-Thermal Resist	SN8-Glosiness
GS9- Spatial Activities	EH9-Nuclear Waste			T9-Moisture Resist	SN9-Hardness
GS10- Material Scale	EH10-Waste Disposal			T10-Scratch Resist	SN10-Lighting Effect
				T11-Weather Resist	SN11-Translucence
				T12-Chemical Resist	SN12-Structure
				T13-Resist Decay	
				T14- Weight	
				T15-Life Expectancy	
				T17-UV Resistance	
				T18-Compatibility	

Alternatives

| Alternative Choice A | Alternative Choice B | Alternative Choice C |

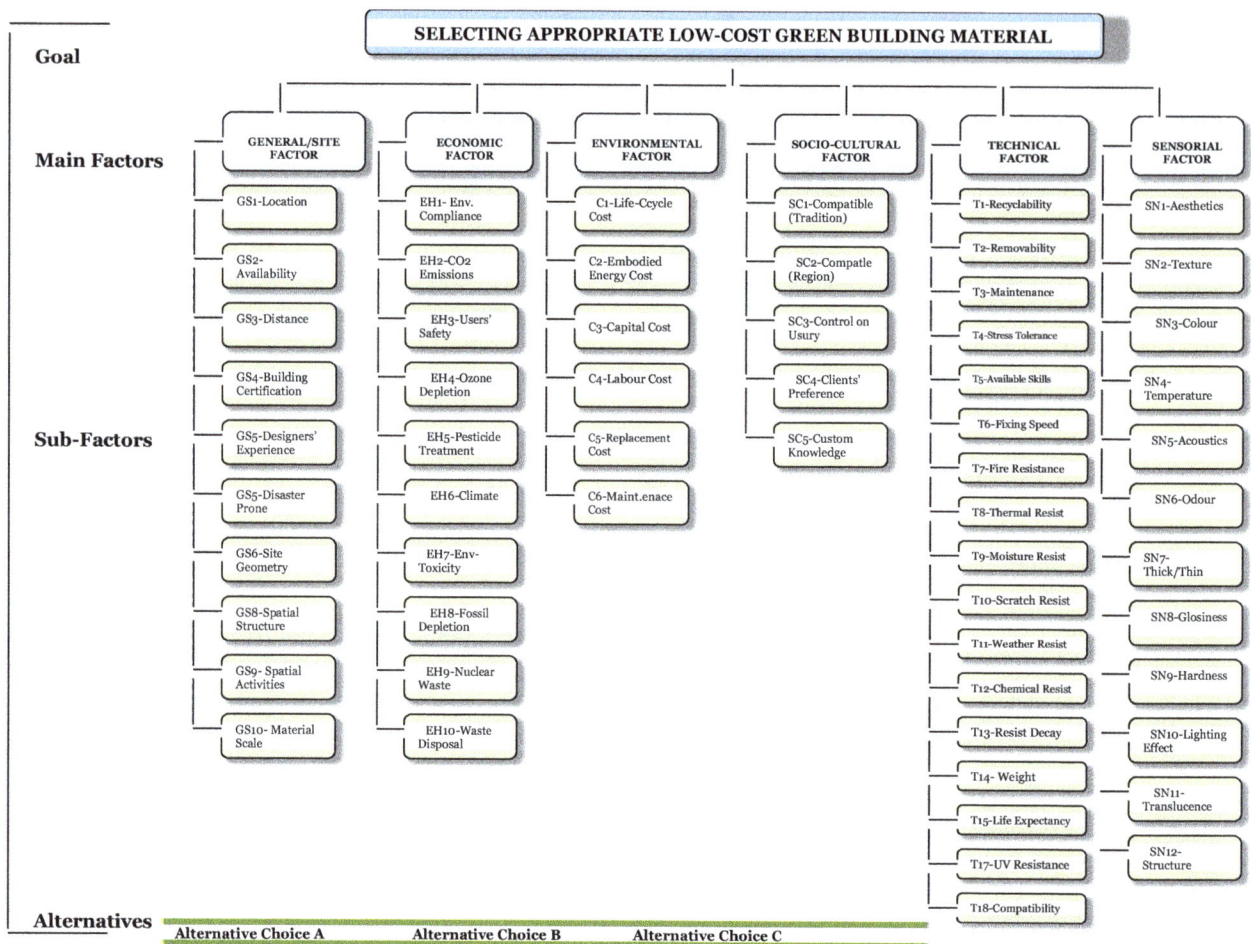

Figure 6. Hierararchy of the material selection phases.

is accepted if the Consistency Ratio (CR) ≤ 0.10. In cases were the results of the respondents were not consistent, the participants were alerted or prompted by the model to carefully re-evaluate the factors until consistency was achieved.

Figures 9 and **10** represent the principal matrix of comparison, which contains the comparison between main/parent factors in relation to the overall objective of the problem (*i.e.*, the selection of a sustainable low-cost green building floor material). From **Figure 9**, it is possible to observe that factor SC is 3 times more important than factor EH. As a logical consequence, factor EH is 3 times less important than factor SC. It is also possible to observe that the elements in the principal diagonal are always equal to 1. In other words, the weight of a criterion in relation to itself, obviously, is always 1.

From **Figure 9**, it is also possible to observe that comparing Socio-cultural [SC] and Technical [T] factors, the participants slightly favoured Technical aspects of the products [T], thus arrived at an average value of two (2), derived from the mean calculation of the five respondents. Comparing Socio-cultural [SC] impacts with Sen-

sorial [SN], participants somewhat considered Socio-cultural [SC] as more relevant in their choice of materials than the emotive or sensorial [SN] aspects of the products, thus arriving at a mean score of 2. Comparing Technical [T] and Sensorial [SN], Technical [T] issues where proven to be more relevant or more slightly favoured than others making it the most dominant factor of the three. Based on their preference values, the system automatically creates a reciprocal matrix on the opposite end as the case may be.

At this stage (as shown in **Figure 11**), ratio scales are defined for pair-wise comparison of the main or parent factors using the ratio scale of 1 - 9. As mentioned earlier, the decision makers obtained values for each parent factor based on their aprioristic knowledge and individual weighting preference. Here, the AHP main criteria matrix is then automatically developed by comparing the relative importance of one parent factor over the other as shown above in **Figure 11**.

Next, the parent criteria matrices are normalised (by dividing a cell value by the sum of each column) and then checked for consistency using Eigen values as

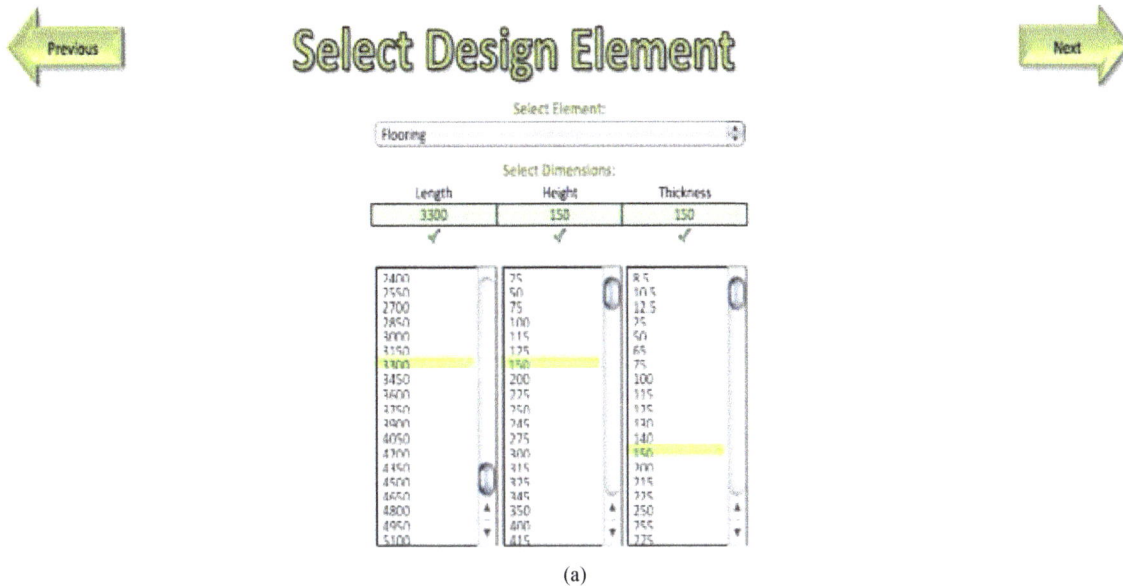

(a)

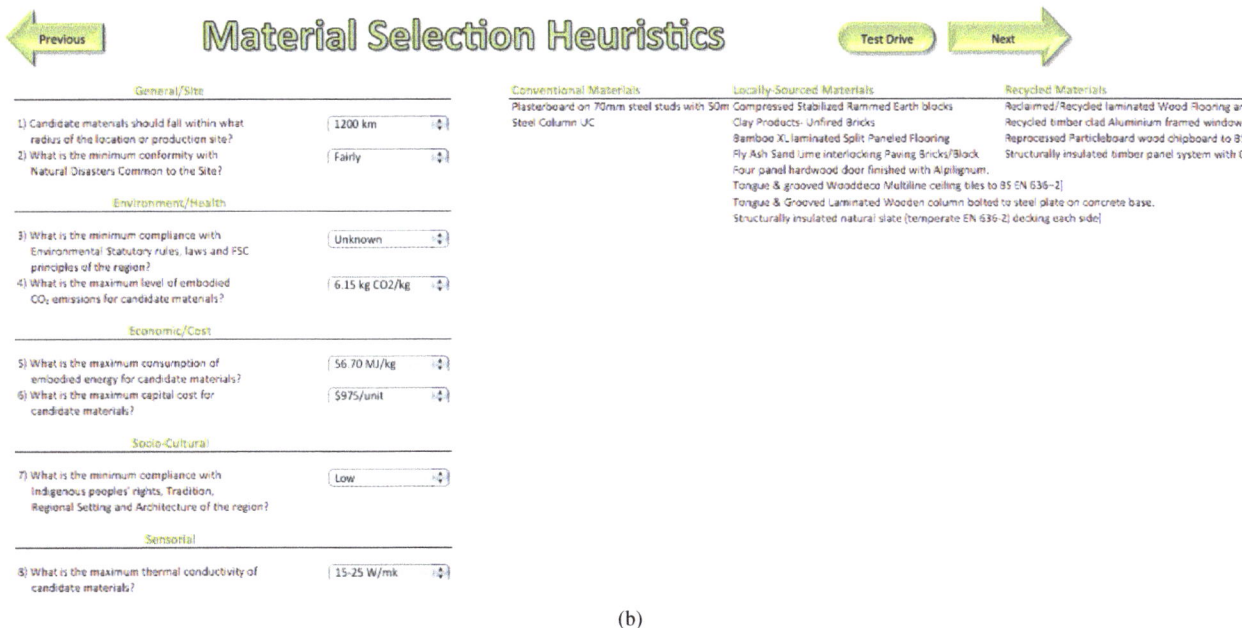

(b)

Figure 7. (a) Dimensional scale for the elected design element; (b) Selection rules for the elected design element.

Ratio Scale For Pairwise Comparisons	
Value (W)	**Definition**
1	Equal Importance of elements
3	Weak Importance of one element over the Other
5	Strong Importance of one element over the other
7	Very Strong Importance of one element over the other
9	Absolute Importance of one element over the other
2,4,6,8	Intermediate values between two adjacent judgements

Figure 8. Ratio scale for pair-wise comparison of factors.

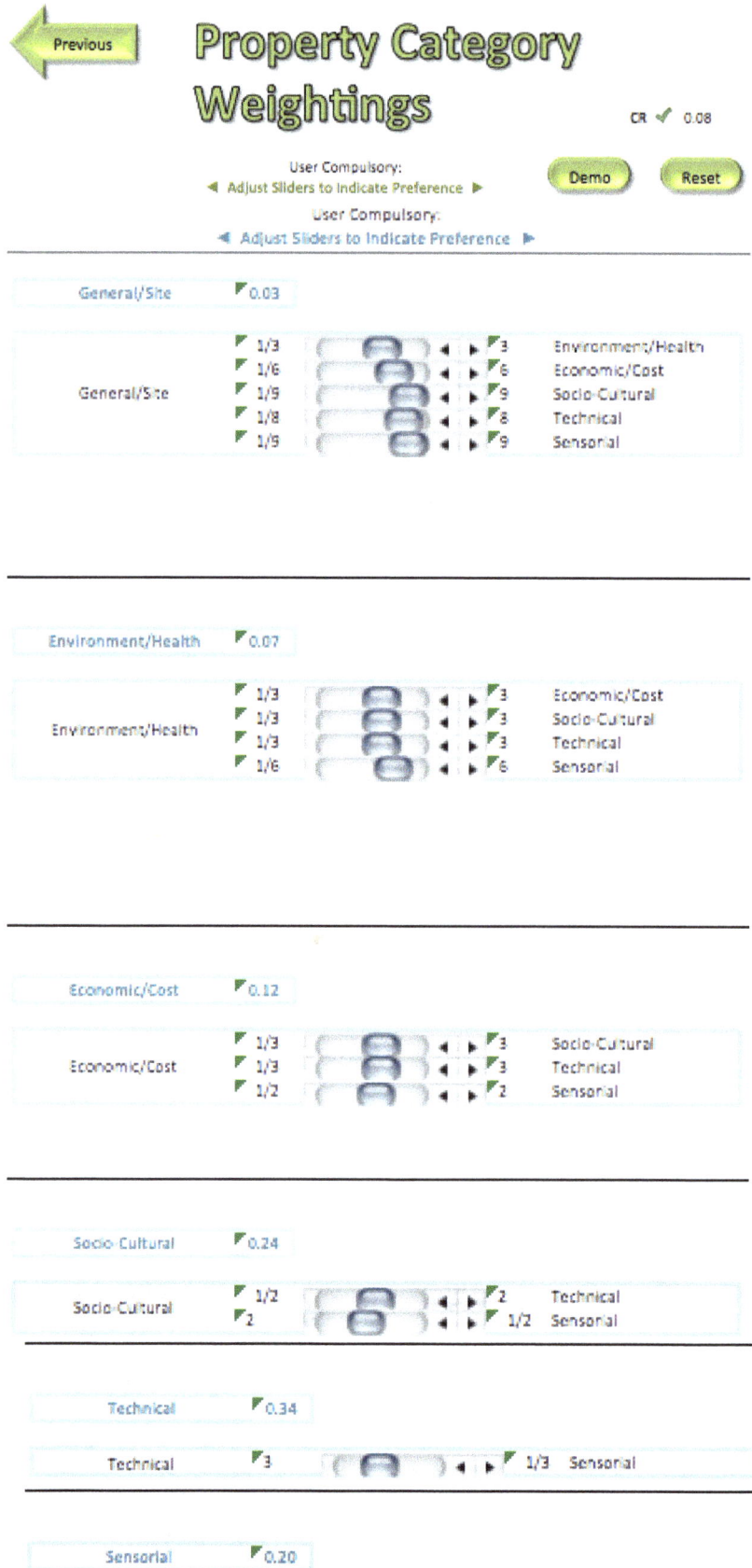

Figure 9. Consensus pair-wise comparison of main factors.

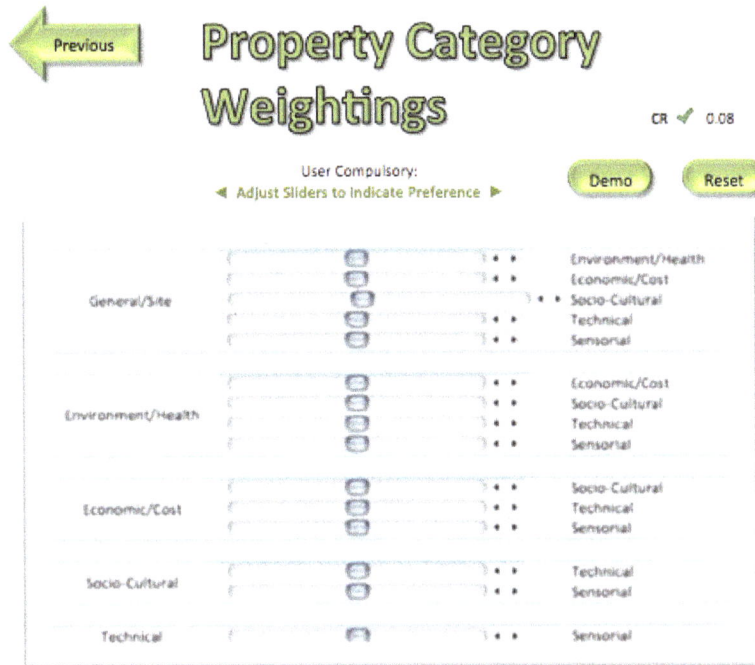

Figure 10. Consensus pair-wise comparison of main factors.

Weighted Criteria Matrix

	General/Site	Environment	Economic/Cost	Socio-Cultural	Technical	Sensorial
General/Site	1.00	0.33	0.17	0.11	0.13	0.11
Environment/Health	3.00	1.00	0.33	0.33	0.33	0.17
Economic/Cost	6.00	3.00	1.00	0.33	0.33	0.50
Socio-Cultural	9.00	3.00	3.00	1.00	0.50	2.00
Technical	8.00	3.00	3.00	2.00	1.00	3.00
Sensorial	9.00	6.00	2.00	0.50	0.33	1.00
Total	36.00	16.33	9.50	4.28	2.63	6.78

Figure 11. Results of pair-wise analysis of parent factors.

shown in **Figure 12**. A local priority vector score is then generated for the matrix of judgments by normalizing the vector in each column of the matrix (*i.e.* dividing each entry of the column by the column total) and then averaging over the rows of the resulting matrix [55]. The normalized eigenvector shown in **Figure 12** represents the relative importance of each parent criteria.

Based on the calculation in **Figure 11**, the relative priorities of the parent factors in the final selection of a sustainable floor material were calculated as displayed in **Figure 12**. The resulting local priority vectors were

given as: (GS = 0.030, EH = 0.070, C = 0.120, SC = 0.240, T = 0.340, and SN = 0.200) as shown in **Table 5**.

In order to measure the level of consistency of the matrix for the parent factors, the consistency index (CI) was then calculated at 0.103 (see **Figure 11**). The random index (RI) was also taken into consideration and values calculated at this stage of the evaluation exercise. According to Saaty (2008), for matrix of order 6, the RI is 1.24 (see **Table 6**). Given the two values (consisting of both the consistency index (CI = 0.103) and the relative index (RI= 1.24), the CR was then calculated as:

Normalised Average Criteria Matrix								
	General/Site	Environment/Health	Economic/Cost	Socio-Cultural	Technical	Sensorial	Av.	λ_{MAX}
General/Site	0.03	0.02	0.02	0.03	0.05	0.02	0.03	0.934297901
Environment/Health	0.08	0.06	0.04	0.08	0.13	0.02	0.07	1.113775203
Economic/Cost	0.17	0.18	0.11	0.08	0.13	0.07	0.12	1.162609985
Socio-Cultural	0.25	0.18	0.32	0.23	0.19	0.30	0.24	1.04719097
Technical	0.22	0.18	0.32	0.47	0.38	0.44	0.34	0.880596922
Sensorial	0.25	0.37	0.21	0.12	0.13	0.15	0.20	1.377336489
Total	1.00	1.00	1.00	1.00	1.00	1.00	1.00	6.52

Matrix Size	6
RI	1.24
CI	0.103
CR	0.083064516

Figure 12. Relative priority scores of the parent factors.

Table 5. Derived priority scores of the parent factors.

Factor/Criterion	Relative Priority
General/Site	0.030
Environmental/Health	0.070
Economic/Cost	0.120
Socio Cultural	0.240
Technical	0.340
Sensorial	0.200

Table 6. Random index values for $1 \leq n \leq 15$.

n	1	2	3	4	5	6	7	8	9	10	11	12	13	14	15
RI	0	0	0.58	0.9	1.12	1.24	1.32	1.41	1.45	1.49	1.51	1.54	1.55	1.57	1.58

CR = CI/RI = 0.103/1.24 = 0.083 (see **Figure 11**).

According to the AHP model, a matrix is considered as being consistent when the CR is less than 10%. With a Consistency Ratio (CR) of 0.083, the matrix was considered consistent since it was less than 0.1.

5.7. Pair-Wise Analysis of Sub-Factors

The results of the next pair-wise comparison matrices amongst the relative sub-factors are shown from **Figures 13-24**. The same calculations done for the principal matrices of the parent factors were also done for the matri-

ces of the sub-factors. The local priority vector and the consistency ratio for each sub-criterion matrix were also computed and displayed on each corresponding table as fully displayed below.

After comparing each sub-factor according to the user's system of value over other sub-factors, the weightings were obtained to establish each priority weightings in the context of the overall goal: selecting the most sustainable low-cost green floor material. The criteria matrices of each sub-factor were then normalised (by dividing a cell value by the sum of each column) and then checked for consistency as shown in **Figures 13-24**.

5.8. Determining the Weightings of Sub-Factors

The next stage of the assessment process was to find the final weightings of both the parent and sub-factors that will be used subsequently to evaluate the material attributes for sustainable building material selection. To determine the final weightings of the selected factors, the priority vectors (1) of the parent factors are multiplied by the corresponding relative priority vectors of each sub-criterion weighting vectors (2) to obtain the (final) weighting (3) as shown in **Table 7**.

The main/parent factor weighting is derived from users' judgment with respect to a single main criterion. The resultant value of the comparison of each parent factor serves as the priority vector of the main criteria needed for evaluating material attributes. The selected value for

Table 7. Derived final weightings for G-site factors.

Criteria	User Value	Default	CR <0.1	Selected Value	Sub-Criteria	User Value	CR < 0.1	Selected Value	Total = 1.0000
Parent factor/Criteria Weighting (1)					**Sub-Factor/Criteria Weighting (2)**				**Final Weighting (3)**
General/Site	0.03	0.057	0.08	**0.026**	**GS1-Location (Mph)**	**0.197**	0.09	**0.197**	**0.0051**
					GS2-Material Availability	0.158		0.158	0.0041
					GS3-Distance to Market (km/h)	0.127		0.127	0.0033
					GS4-Building Certification code	0.115		0.115	0.0030
					GS6-Withstand site natural disaster	0.083		0.083	0.0022
					GS8-Conforms to site geometry	0.114		0.114	0.0030
					GS9-Conforms to spatial structure	0.069		0.069	0.0018
					GS10-Conforms to all spatial activities	0.053		0.053	0.0014
					GS11-Conforms to design geometry	0.044		0.044	0.0012
					GS12-Mat. Spatial scale/Size (sq./m)	0.040		0.040	0.0010

each parent factor as shown in **Table 7** include: GS = 0.026, EH = 0.068, C = 0.122, SC = 0.245, T = 0.335 and SN = 0.203

The sub-factor weighting is derived from user's judgment with respect to each sub-factor. Some of the selected values that serve as the corresponding relative priority vectors of the general/site variable include: 0.197, 0.158, 0.127, 0.115, 0.083, 0.114, 0.069, 0.053, 0.044, and 0.040 as shown in **Table 7**.

Final weighting is derived from multiplying the selected value of the main criteria-weighting or priority vector by the selected value of the sub-factor priority vector. This entry is obtained as follows: 0.026 × 0.197= 0.005122 (as highlighted in **Table 7**). The same process was applied to the other parent factors of the respective categories. The following steps describe the ways by which the various weighting vectors of each criterion are derived.

5.9. Pair-Wise Comparison of the Selected Material Alternatives against Each Sub-Factor

The final step of the exercise was for the respondents to compare each pair of low-cost green material alternatives with respect to each sub-factor. Here the user evaluates the criteria/factors and material alternatives by comparing them through direct rating, to know which factor is more important; how many times; and which material alternative is better in the context of each factor.

The corresponding weightings were based on the im-

portance that the evaluators attached to the dominance of each material alternative relative to all other alternatives under each sub-criterion. These matrices were also normalized and checked for consistency as shown in **Figures 25-38**.

Figures 25-38 present some results of the analyses, which explain the pair-wise matrix priority weightings and normalisation of the various materials with respect to each sub-criterion.

5.10. Determining the Weightings of Sub-Factors

The next phase, after analysing the pair-wise matrices of the sub-factors against the various low-cost green floor material alternatives was to normalize the priority weights for each pair-wise comparison judgment matrices. Once the normalised matrices of the floor material alternatives and various sub-factors were obtained, the values derived from the analysis were multiplied and summed to obtain the final composite priority weights of all material alternatives, focusing particularly on the three floor materials used in the fourth level of the AHP model of decision-making shown in **Figure 6**.

In this case, the final weighting scores (obtained from multiplying the priorities vectors of the parent criteria with that of individual sub-factors), is further multiplied by the priority vector of each material alternative after the pair-wise comparison against each sub-factor (as shown in **Figure 38**). This resulted in a final composite priority/weighting score of each sub-factor for the three floor material alternatives.

	Score	GS1	GS2	GS3	GS4	GS6	GS8	GS9	GS10	GS11	GS12
GS1-Location (Mph)	0.197	1.00	2.00	3.00	2.00	4.00	2.00	2.00	2.00	3.00	3.00
GS2-Material Availability	0.158	0.50	1.00	2.00	2.00	2.00	3.00	3.00	3.00	2.00	3.00
GS3-Distance to Market (km/h)	0.127	0.33	0.50	1.00	2.00	2.00	2.00	3.00	3.00	3.00	2.00
GS4-Building Certification code	0.115	0.50	0.50	0.50	1.00	2.00	2.00	3.00	2.00	4.00	2.00
GS6-Withstand site natural disaster	0.083	0.25	0.50	0.50	0.50	1.00	2.00	2.00	2.00	2.00	2.00
GS8-Conforms to site geometry	0.114	0.50	0.33	0.50	0.50	0.50	1.00	3.00	7.00	3.00	4.00
GS9-Conforms to spatial structure	0.069	0.50	0.33	0.33	0.33	0.50	0.33	1.00	3.00	3.00	2.00
GS10-Conforms to all spatial activities	0.053	0.50	0.33	0.33	0.50	0.50	0.14	0.33	1.00	2.00	2.00
GS11-Conforms to design geometry	0.044	0.33	0.50	0.33	0.25	0.50	0.33	0.33	0.50	1.00	2.00
GS12-Mat. Spatial scale/Size (sq./m)	0.040	0.33	0.33	0.50	0.50	0.50	0.25	0.50	0.50	0.50	1.00
CR	0.09										

Figure 13. Pair-wise matrix for general/site factors.

Normalised Matrix										λ_{MAX}	λ_{MAX}	11
0.210	0.315	0.333	0.208	0.296	0.153	0.110	0.083	0.127	0.130	0.935	**Matrix Size**	10
0.105	0.157	0.222	0.208	0.148	0.229	0.165	0.125	0.085	0.130	0.999	**CI**	0.14
0.070	0.078	0.111	0.208	0.148	0.153	0.165	0.125	0.127	0.086	1.147	**RI**	1.49
0.105	0.078	0.055	0.104	0.148	0.153	0.165	0.083	0.170	0.086	1.103	**CR**	0.09
0.052	0.078	0.055	0.052	0.074	0.153	0.110	0.083	0.085	0.086	1.123		
0.105	0.052	0.055	0.052	0.037	0.076	0.165	0.291	0.127	0.173	1.486		
0.105	0.052	0.037	0.034	0.037	0.025	0.055	0.12	0.127	0.086	1.248		
0.105	0.052	0.037	0.052	0.037	0.010	0.018	0.041	0.085	0.086	1.265		
0.070	0.078	0.037	0.026	0.037	0.025	0.018	0.020	0.042	0.086	1.042		
0.070	0.052	0.055	0.052	0.037	0.019	0.027	0.020	0.021	0.043	0.920		

Figure 14. Normalised matrix for general/site factors.

	Score	EH1	EH2	EH3	EH4	EH5	EH6	EH7	EH8	EH9	EH10
EH1-Env. Statutory Compliance	0.202	1.00	4.00	3.00	2.00	2.00	3.00	3.00	2.00	2.00	2.00
EH2-Embodied CO_2 Emission ($KgCO_2/m^2$)	0.124	0.25	1.00	2.00	3.00	2.00	2.00	2.00	2.00	3.00	0.50
EH3-Human Toxicity-Users Safety level	0.113	0.33	0.50	1.00	2.00	2.00	2.00	3.00	3.00	3.00	0.50
EH4-Ozone depletion rate	0.086	0.50	0.33	0.50	1.00	2.00	2.00	2.00	2.00	2.00	0.33
EH5-Amt. of Pesticide Treatment (l/m^2)	0.078	0.50	0.50	0.50	0.50	1.00	2.00	3.00	2.00	0.33	0.50
EH6-Complies with the Climate of the region	0.067	0.33	0.50	0.50	0.50	0.50	1.00	2.00	2.00	2.00	0.50
EH7-Env. Toxicity (land, water, Animals)	0.053	0.33	0.50	0.33	0.50	0.33	0.50	1.00	2.00	2.00	0.33
EH8-Fossil fuel/Habitat depletion	0.058	0.50	0.50	0.33	0.50	0.50	0.50	0.50	1.00	4.00	0.25
EH9-Nuclear waste rate	0.057	0.50	0.33	0.33	0.50	3.00	0.50	0.50	0.25	1.00	0.33
EH10-Waste Disposal rate	0.162	0.50	2.00	2.00	3.00	2.00	2.00	3.00	4.00	3.00	1.00
CR	0.10										

Figure 15. Pair-wise matrix for environmental factors.

Normalised Matrix										λ_{MAX}	λ_{MAX}	11
0.210	0.393	0.285	0.148	0.130	0.193	0.15	0.098	0.089	0.32	0.960	**Matrix Size**	10
0.052	0.098	0.190	0.222	0.130	0.129	0.1	0.098	0.134	0.08	1.257	**CI**	0.15
0.070	0.049	0.095	0.148	0.130	0.129	0.15	0.148	0.134	0.08	1.191	**RI**	1.49
0.105	0.032	0.047	0.074	0.130	0.129	0.1	0.098	0.089	0.05	1.162	**CR**	0.10
0.105	0.049	0.047	0.037	0.065	0.129	0.15	0.098	0.014	0.08	1.191		
0.070	0.049	0.047	0.037	0.032	0.064	0.1	0.098	0.089	0.08	1.038		
0.070	0.049	0.031	0.037	0.020	0.032	0.05	0.098	0.089	0.05	1.068		
0.105	0.049	0.031	0.037	0.032	0.032	0.025	0.049	0.179	0.04	1.178		
0.105	0.032	0.031	0.037	0.195	0.032	0.025	0.012	0.044	0.05	1.273		
0.105	0.196	0.190	0.222	0.130	0.129	0.15	0.197	0.134	0.16	1.010		

Figure 16. Normalised matrix for environmental factors.

	Score	C1	C2	C3	C4	C5	C6
C1-Total life-cycle cost ($)	0.347	1.00	2.00	2.00	3.00	5.00	9.00
C2-Material embodied energy cost ($)	0.247	0.50	1.00	2.00	4.00	4.00	3.00
C3-Material capital cost ($)	0.186	0.50	0.50	1.00	2.00	4.00	6.00
C4-Labour/Installation cost ($/sqft)	0.120	0.33	0.25	0.50	1.00	3.00	5.00
C5-Material replacement cost ($)	0.063	0.20	0.25	0.25	0.33	1.00	3.00
C6-Material Maintenance cost ($)	0.037	0.11	0.33	0.17	0.20	0.33	1.00
CR	0.07						

Figure 17. Pair-wise matrix for economic/cost factors.

Normalised Matrix						λ_{MAX}	λ_{MAX}	6
0.378	0.461	0.338	0.284	0.288	0.333	0.919	**Matrix Size**	6
0.18	0.230	0.338	0.379	0.230	0.111	1.069	**CI**	0.09
0.18	0.115	0.169	0.189	0.230	0.222	1.101	**RI**	1.24
0.12	0.057	0.084	0.094	0.173	0.185	1.267	**CR**	0.07
0.075	0.057	0.042	0.031	0.057	0.111	1.086		
0.042	0.076	0.028	0.018	0.019	0.037	1.001		

Figure 18. Normalised matrix for economic/cost factors.

	Score	SC1	SC2	SC3	SC4	SC5
SC1-Material compatibility with traditions	0.164	1.00	2.00	0.33	0.50	2.00
SC2-Material compatibility with region	0.102	0.50	1.00	0.50	0.50	0.33
SC3-Cultural restriction on usury	0.362	3.00	2.00	1.00	2.00	3.00
SC4-Client's preference rating	0.227	2.00	2.00	0.50	1.00	2.00
SC5-Conforms to Knowledge of custom	0.146	0.50	3.00	0.33	0.50	1.00
CR	0.08					

Figure 19. Pair-wise matrix for socio-cultural factors.

Normalised Matrix					λ_{MAX}	λ_{MAX}	5
0.142	0.2	0.125	0.111	0.24	1.147	**Matrix Size**	5
0.071	0.1	0.187	0.111	0.04	1.020	**CI**	0.09
0.428	0.2	0.375	0.444	0.36	0.964	**RI**	1.12
0.285	0.2	0.1875	0.222	0.24	1.022	**CR**	0.08
0.071	0.3	0.125	0.111	0.12	1.213		

Figure 20. Normalised matrix for socio-cultural factors.

	Score	T1	T2	T3	T4	T5	T6	T7	T8	T9	T10	T11	T12	T13	T14	T15	T16	T17	T17
T1-Recyclable	0.09	1.00	2.00	2.00	3.00	0.50	2.00	2.00	0.50	0.50	2.00	3.00	2.00	2.00	2.00	3.00	0.50	0.33	0.50
T2-Ease to remove	0.10	0.50	1.00	0.33	0.33	0.33	3.00	2.00	3.00	0.50	2.00	3.00	2.00	2.00	3.00	2.00	3.00	2.00	2.00
T3- Maintenance level	0.06	0.50	3.00	1.00	1.00	1.00	1.00	1.00	1.00	1.00	1.00	1.00	1.00	1.00	1.00	1.00	1.00	1.00	1.00
T4-Expansion Tolerance	0.06	0.33	3.00	1.00	1.00	1.00	1.00	1.00	1.00	1.00	1.00	1.00	1.00	1.00	1.00	1.00	1.00	1.00	1.00
T5-Conforms to skills	0.06	2.00	3.00	1.00	1.00	1.00	1.00	1.00	1.00	1.00	1.00	1.00	1.00	1.00	1.00	1.00	1.00	1.00	1.00
T6-Ease of fixing	0.05	0.50	0.33	1.00	1.00	1.00	1.00	1.00	1.00	1.00	1.00	1.00	1.00	1.00	1.00	1.00	1.00	1.00	1.00
T7-Fire resistance	0.04	0.50	0.50	1.00	1.00	1.00	1.00	1.00	1.00	1.00	1.00	1.00	1.00	0.11	1.00	0.14	1.00	1.00	1.00
T8-Thermal resistance	0.05	2.00	0.33	1.00	1.00	1.00	1.00	1.00	1.00	1.00	1.00	1.00	1.00	1.00	1.00	1.00	0.14	1.00	1.00
T9-Moisture resistance	0.06	2.00	2.00	1.00	1.00	1.00	1.00	1.00	1.00	1.00	1.00	1.00	1.00	1.00	1.00	1.00	1.00	1.00	1.00
T10-Scratch resistance	0.05	0.50	0.50	1.00	1.00	1.00	1.00	1.00	1.00	1.00	1.00	1.00	1.00	1.00	1.00	1.00	1.00	1.00	1.00
T11-Weather resistance	0.05	0.33	0.33	1.00	1.00	1.00	1.00	1.00	1.00	1.00	1.00	1.00	1.00	1.00	1.00	1.00	1.00	1.00	1.00
T12-Chemical resistance	0.05	0.50	0.50	1.00	1.00	1.00	1.00	1.00	1.00	1.00	1.00	1.00	1.00	1.00	1.00	1.00	1.00	1.00	1.00
T13-Resistance to decay	0.07	0.50	0.50	1.00	1.00	1.00	1.00	9.00	1.00	1.00	1.00	1.00	1.00	1.00	1.00	1.00	1.00	1.00	1.00
T14-Weight of material	0.05	0.50	0.33	1.00	1.00	1.00	1.00	1.00	1.00	1.00	1.00	1.00	1.00	1.00	1.00	1.00	1.00	1.00	1.00
T15-Life expectancy	0.07	0.33	0.50	1.00	1.00	1.00	1.00	7.00	1.00	1.00	1.00	1.00	1.00	1.00	1.00	1.00	0.25	1.00	1.00
T16-Biodegradable	0.08	2.00	0.33	1.00	1.00	1.00	1.00	1.00	7.00	1.00	1.00	1.00	1.00	1.00	1.00	4.00	1.00	1.00	1.00
T17-UV Resistance	0.06	3.00	0.50	1.00	1.00	1.00	1.00	1.00	1.00	1.00	1.00	1.00	1.00	1.00	1.00	1.00	1.00	1.00	1.00
T18-Compatibility	0.05	0.50	1.00	1.00	1.00	1.00	1.00	1.00	1.00	1.00	1.00	1.00	1.00	1.00	1.00	1.00	1.00	1.00	1.00
CR	0.09																		

Figure 21. Pair-wise matrix for technical factors.

Normalised Matrix																			λ_{MAX}	λ_{MAX}	21
0.05	0.11	0.11	0.17	0.02	0.11	0.11	0.02	0.02	0.11	0.17	0.11	0.11	0.11	0.17	0.02	0.01	0.02	1.602	**Size**	18
0.02	0.05	0.01	0.01	0.01	0.17	0.11	0.17	0.02	0.11	0.17	0.11	0.11	0.17	0.11	0.17	0.11	0.11	1.778	**CI**	0.15
0.02	0.17	0.05	0.05	0.05	0.05	0.05	0.05	0.05	0.05	0.05	0.05	0.05	0.05	0.05	0.05	0.05	0.05	1.083	**RI**	1.69
0.01	0.17	0.05	0.05	0.05	0.05	0.05	0.05	0.05	0.05	0.05	0.05	0.05	0.05	0.05	0.05	0.05	0.05	1.074	**CR**	0.09
0.11	0.17	0.05	0.05	0.05	0.05	0.05	0.05	0.05	0.05	0.05	0.05	0.05	0.05	0.05	0.05	0.05	0.05	1.167		
0.02	0.01	0.05	0.05	0.05	0.05	0.05	0.05	0.05	0.05	0.05	0.05	0.05	0.05	0.05	0.05	0.05	0.05	0.935		
0.02	0.02	0.05	0.05	0.05	0.05	0.05	0.05	0.05	0.05	0.05	0.05	0.00	0.05	0.01	0.05	0.05	0.05	0.847		
0.11	0.01	0.05	0.05	0.05	0.05	0.05	0.05	0.05	0.05	0.05	0.05	0.05	0.05	0.05	0.01	0.05	0.05	0.971		
0.11	0.11	0.05	0.05	0.05	0.05	0.05	0.05	0.05	0.05	0.05	0.05	0.05	0.05	0.05	0.05	0.05	0.05	1.111		
0.02	0.02	0.05	0.05	0.05	0.05	0.05	0.05	0.05	0.05	0.05	0.05	0.05	0.05	0.05	0.05	0.05	0.05	0.944		
0.01	0.01	0.05	0.05	0.05	0.05	0.05	0.05	0.05	0.05	0.05	0.05	0.05	0.05	0.05	0.05	0.05	0.05	0.926		
0.02	0.02	0.05	0.05	0.05	0.05	0.05	0.05	0.05	0.05	0.05	0.05	0.05	0.05	0.05	0.05	0.05	0.05	0.944		
0.02	0.02	0.05	0.05	0.05	0.05	0.51	0.05	0.05	0.05	0.05	0.05	0.05	0.05	0.05	0.05	0.05	0.05	1.389		
0.02	0.01	0.05	0.05	0.05	0.05	0.05	0.05	0.05	0.05	0.05	0.05	0.05	0.05	0.05	0.05	0.05	0.05	0.935		
0.01	0.02	0.05	0.05	0.05	0.05	0.4	0.05	0.05	0.05	0.05	0.05	0.05	0.05	0.05	0.01	0.05	0.05	1.227		
0.11	0.01	0.05	0.05	0.05	0.05	0.05	0.4	0.05	0.05	0.05	0.05	0.05	0.05	0.22	0.05	0.05	0.05	1.519		
0.17	0.02	0.05	0.05	0.05	0.05	0.05	0.05	0.05	0.05	0.05	0.05	0.05	0.05	0.05	0.05	0.05	0.05	1.083		
0.02	0.05	0.05	0.05	0.05	0.05	0.05	0.05	0.05	0.05	0.05	0.05	0.05	0.05	0.05	0.05	0.05	0.05	0.972		

Figure 22. Normalised matrix for technical factors.

	Score	SN1	SN2	SN3	SN4	SN5	SN6	SN7	SN8	SN9	SN10	SN11	SN12	SN13
SN1-Aesthetics	0.077	1.00	1	1	1	1	1	1	1	1	1	1	1	1
SN2-Texture	0.077	1.00	1.00	1	1	1	1	1	1	1	1	1	1	1
SN3-Colour	0.077	1.00	1.00	1.00	1	1	1	1	1	1	1	1	1	1
SN4-Temperature	0.077	1.00	1.00	1.00	1.00	1	1	1	1	1	1	1	1	1
SN5-Acoustics	0.106	1.00	1.00	1.00	1.00	1.00	2	0	4	0	2	0	2	2
SN6-Odour	0.087	1.00	1.00	1.00	1.00	0.50	1.00	2	1	0	2	1	2	2
SN7-Thickness/Thinness	0.107	1.00	1.00	1.00	1.00	3.00	0.50	1.00	2	2	2	3	0	0
SN8-Glossiness/fineness	0.075	1.00	1.00	1.00	1.00	0.25	2.00	0.50	1.00	1	1	1	1	1
SN9-Strength/Hardness	0.109	1.00	1.00	1.00	1.00	3.00	5.00	0.50	1.00	1.00	1	1	1	1
SN10-Lighting effect	0.068	1.00	1.00	1.00	1.00	0.50	0.50	0.50	1.00	1.00	1.00	1	1	1
SN11-Translucence	0.108	1.00	1.00	1.00	1.00	6.00	2.00	0.33	1.00	1.00	1.00	1.00	1	1
SN12-Structure	0.089	1.00	1.00	1.00	1.00	0.50	0.50	4.00	1.00	1.00	1.00	1.00	1.00	1
SN13-Thermal	0.083	1.00	1.00	1.00	1.00	0.50	0.50	3.00	1.00	1.00	1.00	1.00	1.00	1.00
CR	0.10													

Figure 23. Pair-wise matrix for sensorial factors.

Normalised Matrix													λ_{MAX}
0.076	0.076	0.076	0.076	0.076	0.076	0.076	0.076	0.076	0.076	0.076	0.076	0.076	1.000
0.076	0.076	0.076	0.076	0.076	0.076	0.076	0.076	0.076	0.076	0.076	0.076	0.076	1.000
0.076	0.076	0.076	0.076	0.076	0.076	0.076	0.076	0.076	0.076	0.076	0.076	0.076	1.000
0.076	0.076	0.076	0.076	0.076	0.076	0.076	0.076	0.076	0.076	0.076	0.076	0.076	1.000
0.076	0.076	0.076	0.076	0.076	0.153	0.025	0.307	0.025	0.153	0.012	0.153	0.153	1.372
0.076	0.076	0.076	0.076	0.038	0.076	0.153	0.038	0.015	0.153	0.038	0.153	0.153	1.131
0.076	0.076	0.076	0.076	0.230	0.038	0.076	0.153	0.153	0.153	0.230	0.019	0.025	1.391
0.076	0.076	0.076	0.076	0.019	0.153	0.038	0.076	0.076	0.076	0.076	0.076	0.076	0.981
0.076	0.076	0.076	0.076	0.230	0.384	0.038	0.076	0.076	0.076	0.076	0.076	0.076	1.423
0.076	0.076	0.076	0.076	0.038	0.038	0.038	0.076	0.076	0.076	0.076	0.076	0.076	0.885
0.076	0.076	0.076	0.076	0.461	0.153	0.025	0.076	0.076	0.076	0.076	0.076	0.076	1.410
0.076	0.076	0.076	0.076	0.038	0.038	0.307	0.076	0.076	0.076	0.076	0.076	0.076	1.154
0.076	0.076	0.076	0.076	0.038	0.038	0.230	0.076	0.076	0.076	0.076	0.076	0.076	1.077

λ_{MAX}		15
Matrix Size		13
CI		0.15
RI		1.5551
CR		0.10

Figure 24. Normalised matrix for sensorial factors.

Using the priorities determined through these matrices, the weighted overall priority of each candidate material was determined. The amalgamation method yielded a single green utility index of alternative worth, which allowed the material options to be ranked according to their overall priorities. The material with the highest score then becomes the selected candidate material as shown in **Figure 38**. Looking at **Figure 38**, it is clear from the results of the analysis that Material option (A) turns out to be the most preferred material among the three material options identified in **Table 4**, with an overall priority or index score of 0.086. It is based on the concept of the higher the green utility index value, the better the option. The green utility index as calculated for each of the three material alternatives was M(C) = 0.086, M(A) = 0.072 and M(B) = 0.062 for material options C,

A and B respectively, making Option C (fly-ash cement concrete floor slab) emerge as the best option amongst the other alternatives as shown in **Figure 38**.

The above example has illustrated the application of the MSDSS in a material selection problem for a proposed 5-bedroom low-cost residential green building project in the London Borough of Sutton. From the illustrated example it can be deduced that the MSDSS model is able to provide rankings in low-cost green building material assessment combining site, economic, technical, social-cultural, sensorial and environmental criteria into a composite index system based on the AHP technique. This model is therefore, based on the presumption that decision makers, given full knowledge of all possible consequences of all possible alternatives and factors, will select the material with the highest-ranking score.

GS1-Location (km)	CSR	CP	RL	B.XL	FA	RT	FPH.	SS	RPB	T&GW	PB	T&G	SC	SIT	
Compressed Stabilized Rammed Earth blocks	1.0	2.0	2.0	4.0	2.0	5.0	8.0	8.0	4.0	4.00	4.0	4.00	7.0	2.00	4.0
Clay Products-Unfired Bricks	0.5	1.0	1.0	3.0	1.0	4.0	7.0	7.0	3.0	3.00	3.0	3.00	6.0	1.00	3.0
Reclaimed/Recycled laminated Wood Flooring and Panelling	0.5	1.0	1.0	3.0	1.0	4.0	7.0	7.0	3.0	3.00	3.0	3.00	6.0	1.00	3.0
Bamboo XL laminated Split Paneled Flooring	0.3	0.3	0.3	1.0	0.3	2.0	5.0	5.0	1.0	1.0	1.0	1.00	4.0	0.33	1.0
Fly Ash Sand Lime interlocking Paving Bricks/Block	0.5	1.0	1.0	3.0	1.0	4.00	7.00	7.00	3.0	3.00	3.0	3.0	6.0	1.0	3.0
Recycled timber clad Aluminium framed window unit	0.2	0.3	0.3	0.5	0.3	1.0	4.00	4.00	0.50	0.50	0.5	0.50	3.0	0.3	0.5
Four panel hardwood door finished with Alpilignum.	0.1	0.1	0.1	0.2	0.1	0.3	1.0	1.0	0.2	0.2	0.2	0.2	0.5	0.1	0.2
Stainless Steel Entry Door.	0.1	0.1	0.1	0.2	0.1	0.3	1.0	1.0	0.2	0.2	0.2	0.2	0.5	0.1	0.20
Reprocessed Particleboard wood chipboard to BS EN 312 Type P5,	0.3	0.3	0.3	1.0	0.3	2.0	5.0	5.0	1.0	1.00	1.0	1.00	4.0	0.3	1.00
Tongue & grooved Wooddeco Multiline ceiling tiles to BS EN 636–2|	0.3	0.3	0.3	1.0	0.3	2.0	5.0	5.0	1.0	1.00	1.0	1.00	4.0	0.33	1.00
Plasterboard on 70 mm steel studs with 50 mm 12.9 kg/m^3 insulation,	0.3	0.3	0.3	1.0	0.3	2.0	5.0	5.0	1.0	1.00	1.0	1.00	4.00	0.33	1.00
Tongue & Grooved Laminated Wooden column bolted to steel plate on concrete base.	0.3	0.3	0.3	1.0	0.3	2.0	5.0	5.0	1.0	1.0	1.0	1.00	4.0	0.33	1.0
Steel Column UC	0.1	0.2	0.2	0.3	0.2	0.33	2.00	2.00	0.3	0.25	0.3	0.3	1.0	0.17	0.3
Structurally insulated timber panel system with OSB/3 each side, roofing underlay reclaimed clay tiles	0.5	1.0	1.0	3.0	1.0	4.0	7.0	7.0	3.0	3.0	3.0	3.0	6.0	1.0	3.0
Structurally insulated natural slate (temperate EN 636-2) decking each side|	0.3	0.3	0.3	1.0	0.3	2.0	5.00	5.00	1.00	1.00	1.0	1.00	4.0	0.3	1.0
Total	5.1	8.7	8.7	23.2	8.7	34.8	74.0	74.0	23.2	23.2	23.2	23.2	60.0	8.7	23.2

Figure 25. Pair-wise matrix: location.

CS	CP	RL	B.XL	FA	RT	FPH.	SS.	RP,	T&G]	PB	T&GW.	SC	SIT	SIS
0.2	0.2	0.2	0.2	0.23	0.14	0.11	0.11	0.17	0.17	0.17	0.17	0.12	0.23	0.17
0.1	0.1	0.1	0.1	0.11	0.11	0.09	0.09	0.13	0.13	0.13	0.13	0.10	0.11	0.13
0.1	0.1	0.1	0.1	0.11	0.11	0.09	0.09	0.13	0.13	0.13	0.13	0.10	0.11	0.13
0.0	0.0	0.0	0.0	0.04	0.1	0.07	0.07	0.04	0.04	0.04	0.04	0.07	0.04	0.04
0.10	0.11	0.11	0.13	0.11	0.11	0.09	9.46E-02	0.13	0.13	0.13	0.13	0.10	0.11	0.13
0.0	0.0	0.03	0.02	0.03	0.03	0.05	0.05	0.02	0.02	0.02	0.02	0.05	0.03	0.02
0.0	0.0	0.0	0.0	0.02	0.0	0.01	0.0135134	0.01	0.01	0.01	0.01	0.01	0.02	0.01
0.0	0.0	0.0	0.0	0.02	0.0	0.01	4	0.01	0.01	0.01	0.01	0.01	0.02	0.01
0.0	0.0	0.0	0.0	0.04	0.1	0.07	0.067567568	0.04	0.04	0.04	0.04	0.07	0.04	0.04
0.0	0.0	0.0	0.0	0.0	0.1	0.1	0.067567568	0.04	0.04	0.04	0.04	0.07	0.04	0.04
0.05	0.04	0.04	0.0	0.04	0.1	0.1	0.067567568	0.04	0.04	0.04	0.04	0.07	0.04	0.04
0.0	0.0	0.0	0.0	0.0	0.1	0.1	0.067567568	0.04	0.04	0.04	0.04	0.07	0.04	0.04
0.0	0.0	0.0	0.0	0.0	0.0	0.0	0.027027027	0.01	0.01	0.01	0.01	0.02	0.02	0.01
0.1	0.1	0.1	0.1	0.1	0.1	0.1	0.094594595	0.13	0.13	0.13	0.13	0.10	0.11	0.13
0.0	0.0	0.0	0.0	0.04	0.1	0.07	0.067567568	0.04	0.04	0.04	0.04	0.07	0.04	0.04
1.00	1.00	1.00	1.00	1.00	1.00	1.00	1.00	1.00	1.00	1.00	1.00	1.00	1.00	1.00

Figure 26. Normalised matrix: location.

EH2-Embodied CO_2 Emission ($KgCO_2/m^2$)	Compressed Stabilized Rammed Earth blocks	Clay Products- Unfired Bricks	Reclaimed/Recycled laminated Wood Flooring and Panelling	Bamboo XL laminated Split Paneled Flooring	Fly Ash Sand Lime interlocking Paving Bricks/Block	Recycled timber clad Aluminium framed window unit	Four panel hardwood door finished with Alpilignum.	Stainless Steel Entry Door.	Reprocessed Particleboard wood chipboard to BS EN 312 Type P5,	Tongue & grooved Wooddeco Multiline ceiling tiles to BS EN 636-2]	Plasterboard on 70 mm steel studs with 50 mm 12.9 kg/m³ insulation,	Tongue & Grooved Laminated Wooden column bolted to steel plate on concrete base.	Steel Column UC	Structurally insulated timber panel system with OSB/3 each side, roofing underlay reclaimed clay tiles	Structurally insulated natural slate (temperate EN 636-2) decking each side]
Compressed Stabilized Rammed Earth blocks	1.0	1.0	5.0	1.0	1.0	5.0	5.0	8.0	2.0	1.0	4.0	5.00	6.00	5.00	1.00
Clay Products—Unfired Bricks	1.0	1.0	5.0	1.0	1.0	5.0	5.0	8.0	2.0	1.0	4.0	5.0	6.0	5.0	1.0
Reclaimed/Recycled laminated Wood Flooring and Panelling	0.2	0.2	1.0	0.2	0.2	1.0	1.0	4.0	0.3	0.2	0.5	1.0	2.0	1.0	0.2
Bamboo XL laminated Split Paneled Flooring	1.0	1.0	5.0	1.0	1.0	5.0	5.0	8.0	2.0	1.0	4.0	5.0	6.0	5.0	1.0
Fly Ash Sand Lime interlocking Paving Bricks/Block	1.0	1.0	5.0	1.0	1.0	5.0	5.0	8.0	2.0	1.0	4.0	5.0	6.0	5.0	1.0
Recycled timber clad Aluminium framed window unit	0.2	0.2	1.0	0.2	0.2	1.0	1.0	4.0	0.3	0.2	0.5	1.0	2.0	1.0	0.2
Four panel hardwood door finished with Alpilignum.	0.2	0.2	1.0	0.2	0.2	1.0	1.0	4.0	0.3	0.2	0.5	1.0	2.0	1.0	0.2
Stainless Steel Entry Door.	0.125	0.125	0.25	0.125	0.125	0.25	0.25	1	0.14	0.125	0.2	0.25	0.3	0.25	0.125
Reprocessed Particleboard wood chipboard to BS EN 312 Type P5,	0.5	0.5	4.0	0.5	0.5	4.0	4.0	7.0	1.0	0.5	3.0	4.0	5.0	4.0	0.5
Tongue & grooved Wooddeco Multiline ceiling tiles to BS EN 636–2]	1	1	5	1	1	5	5	8	2	1	4	5	6	5	1
Plasterboard on 70 mm steel studs with 50 mm 12.9 kg/m³ insulation,	0.25	0.25	2	0.25	0.25	2	2	5	0.3	0.25	1	2	3	2	0.25
Tongue & Grooved Laminated Wooden column bolted to steel plate on concrete base.	0.20	0.20	1.00	0.20	0.20	1.00	1.00	4.00	0.25	0.20	0.50	1.00	2.00	1.00	0.20
Steel Column UC	0.2	0.2	0.5	0.2	0.2	0.5	0.5	3.0	0.2	0.2	0.3	0.5	1.0	0.5	0.2
Structurally insulated timber panel system with OSB/3 each side, roofing underlay reclaimed clay tiles	0.2	0.2	1.0	0.2	0.2	1.0	1.0	4.0	0.3	0.2	0.5	1.0	2.0	1.0	0.2
Structurally insulated natural slate (temperate EN 636-2) decking each side]	1.0	1.0	5.0	1.0	1.0	5.0	5.0	8.0	2.0	1.0	4.0	5.00	6.00	5.00	1.00
Total			8.0		8.0		41.8	8.0	8.0	41.8	41.8	84.0	14.9		8.0

Figure 27. Pair-wise matrix: embodied CO_2 emissions.

Compressed Stabilized Rammed Earth blocks	Clay Products- Unfired Bricks	Reclaimed/Recycled laminated Wood Flooring and Panelling	Bamboo XL laminated Split Paneled Flooring	Fly Ash Sand Lime interlocking Paving Bricks/Block	Recycled timber clad Aluminium framed window unit	Four panel hardwood door finished with Alpilignum.	Stainless Steel Entry Door.	Reprocessed Particleboard wood chipboard to BS EN 312 Type P5,	Tongue & grooved Wooddeco Multiline ceiling tiles to BS EN 636-2]	Plasterboard on 70 mm steel studs with 50 mm 12.9 kg/m³ insulation,	Tongue & Grooved Laminated Wooden column bolted to steel plate on concrete base.	Steel Column UC	Structurally insulated timber panel system with OSB/3 each side, roofing underlay reclaimed clay tiles	Structurally insulated natural slate (temperate EN 636-2) decking each side]	CI	0.03	
0.12	0.12	0.12	0.1	0.12	0.1	0.1	0.1	0.12	0.12	0.13	0.12	0.11	0.12	0.12	0.12	0.97	RI 1.58
0.1	0.1	0.1	0.1	0.1	0.1	0.1	0.1	0.1	0.12	0.13	0.12	0.11	0.12	0.12	0.12	0.97	CR 0.02
0.0	0.0	0.0	0.0	0.0	0.0	0.0	0.05	0.01	0.02	0.02	0.02	0.04	0.02	0.02	0.03	1.07	
0.1	0.1	0.1	0.1	0.1	0.1	0.1	0.1	0.12	0.12	0.13	0.12	0.11	0.12	0.12	0.12	0.97	
0.1	0.1	0.1	0.1	0.12	0.1	0.12	0.1	0.12	0.12	0.13	0.12	0.11	0.12	0.12	0.12	0.97	
0.0	0.0	0.0	0.0	0.02	0.0	0.02	0.05	0.01	0.02	0.02	0.02	0.04	0.02	0.02	0.03	1.07	
0.0	0.0	0.0	0.0	0.02	0.0	0.02	0.05	0.01	0.02	0.02	0.02	0.04	0.02	0.02	0.03	1.07	
0.015544041	0.015544041	0.005988024	0.01	0.02	0.004	0.01	0.01	0.009	0.015544041	0.03	0.005	0.006	0.005	0.015	0.01	0.88	
0.1	0.1	0.1	0.1	0.06	0.1	0.10	0.1	0.06	0.06	0.10	0.10	0.09	0.10	0.06	0.08	1.18	
0.124352332	0.124352332	0.119760479	0.12	0.12	0.11	0.12	0.1	0.1	0.124352332	0.12	0.11	0.105	0.11	0.122	0.12	0.97	
0.031088083	0.031088083	0.047904192	0.03	0.03	0.04	0.05	0.1	0.021	0.031088083	0.03	0.047	0.057	0.047	0.03	0.04	1.23	
0.02	0.02	0.02	0.02	0.02	0.02	0.02	0.1	0.01	0.024870466	0.01	0.02	0.038	0.02	0.026	0.03	1.07	
0.0	0.0	0.0	0.0	0.02	0.0	0.01	0.01	0.01	0.02	0.01	0.01	0.02	0.01	0.02	0.02	0.97	
0.0	0.0	0.0	0.0	0.0	0.0	0.0	0.05	0.01	0.02	0.02	0.02	0.04	0.02	0.02	0.03	1.07	
0.12	0.12	0.12	0.1	0.12	0.1	0.1	01	0.1	0.12	0.13	0.12	0.11	0.12	0.12	0.12	0.97	
1.00	1.00	1.00	1.00	1.00	1.00	1.00	1.00	1.00	1.00	1.00	1.00	1.00	1.00	1.00	1.00	15.5	

Figure 28. Normalised matrix: embodied CO₂ emissions.

C1- Total life-cycle cost ($)	Compressed Stabilized Rammed Earth blocks	Clay Products- Unfired Bricks	Reclaimed/Recycled laminated Wood Flooring and Panelling	Bamboo XL laminated Split Paneled Flooring	Fly Ash Sand Lime interlocking Paving Bricks/Block	Recycled timber clad Aluminium framed window unit	Four panel hardwood door finished with Alpilignum.	Stainless Steel Entry Door.	Reprocessed Particleboard wood chipboard to BS EN 312 Type P5,	Tongue & grooved Wooddeco Multiline ceiling tiles to BS EN 636–2]	Plasterboard on 70 mm steel studs with 50 mm 12.9 kg/m³ insulation,	Tongue & Grooved Laminated Wooden column bolted to steel plate on concrete base.	Steel Column UC	Structurally insulated timber panel system with OSB/3 each side, roofing underlay reclaimed clay tiles	Structurally insulated natural slate (temperate EN 636-2) decking each side]
Compressed Stabilized Rammed Earth blocks	1.0	0.5	3.0	0.5	2.0	7.0	8.0	7.0	7.0	8.0	8.0	8.0	7.0	7.0	7.0
Clay Products- Unfired Bricks	2	1	4	1	3	8	9	8	8	9	9	9	8	8	8
Reclaimed/Recycled laminated Wood Flooring and Panelling	0.3	0.3	1.0	0.3	0.5	5.0	6.0	5.0	5.0	6.0	6.0	6.00	5.00	5.00	5.00
Bamboo XL laminated Split Paneled Flooring	2	1	4	1	3	8	9	8	8	9	9	9	8	8	8
Fly Ash Sand Lime interlocking Paving Bricks/Block	0.5	0.3	2	0.3	1	6	7	6	6	7	7	7	6	6	6
Recycled timber clad Aluminium framed window unit	0.14	0.13	0.20	0.13	0.17	1.00	2.00	1.00	1.00	2.00	2.00	2.00	1.00	1.00	1.00
Four panel hardwood door finished with Alpilignum.	0.1	0.1	0.2	0.1	0.1	0.5	1.0	0.5	0.5	1.0	1.0	1.0	0.5	0.5	0.5
Stainless Steel Entry Door.	0.1	0.1	0.2	0.1	0.2	1.0	2.0	1.0	1.0	2.0	2.0	2.0	1.0	1.0	1.0
Reprocessed Particleboard wood chipboard to BS EN 312 Type P5,	0.1	0.1	0.2	0.1	0.2	1.0	2.0	1.0	1.0	2.0	2.0	2.0	1.0	1.0	1.0
Tongue & grooved Wooddeco Multiline ceiling tiles to BS EN 636–2]	0.1	0.1	0.2	0.1	0.1	0.5	1.0	0.5	0.5	1.0	1.0	1.0	0.5	0.5	0.5
Plasterboard on 70 mm steel studs with 50 mm 12.9 kg/m³ insulation,	0.1	0.1	0.2	0.1	0.1	0.5	1.0	0.5	0.5	1.0	1.0	1.0	0.5	0.5	0.5
Tongue & Grooved Laminated Wooden column bolted to steel plate on concrete base.	0.1	0.1	0.2	0.1	0.1	0.5	1.0	0.5	0.5	1.0	1.0	1.0	0.5	0.5	0.5
Steel Column UC	0.1	0.1	0.2	0.1	0.2	1.0	2.0	1.0	1.0	2.0	2.0	2.0	1.0	1.0	1.0
Structurally insulated timber panel system with OSB/3 each side, roofing underlay reclaimed clay tiles	0.1	0.1	0.2	0.1	0.2	1.0	2.0	1.0	1.0	2.0	2.0	2.0	1.0	1.0	1.0
Structurally insulated natural slate (temperate EN 636-2) decking each side]	0.1	0.1	0.2	0.1	0.2	1.0	2.0	1.0	1.0	2.0	2.0	2.0	1.0	1.0	1.0
Total	7.2	4.3	15.9	4.3	11.1	42.0	55.0	42.0	42.0	55.0	55.0	55.0	42.0	42.0	42.0

Figure 29. Pair-wise matrix: total life-cycle cost.

Compressed Stabilized Rammed Earth blocks	Clay Products- Unfired Bricks	Reclaimed/Recycled laminated Wood Flooring and Panelling	Bamboo XL laminated Split Paneled Flooring	Fly Ash Sand Lime interlocking Paving Bricks/Block	Recycled timber clad Aluminium framed window unit	Four panel hardwood door finished with Alpilignum.	Stainless Steel Entry Door.	Reprocessed Particleboard wood chipboard to BS EN 312 Type P5,	Tongue & grooved Wooddeco Multiline ceiling tiles to BS EN 636–2]	Plasterboard on 70 mm steel studs with 50 mm 12.9 kg/m³ insulation,	Tongue & Grooved Laminated Wooden column bolted to steel plate on concrete base.	Steel Column UC	Structurally insulated timber panel system with OSB/3 each side, roofing underlay reclaimed clay tiles	Structurally insulated natural slate (temperate EN 636–2) decking each side]	Average	Lambda Max	CI	0.06
0.1	0.1	0.2	0.1	0.18	0.2	0.15	0.17	0.1667	0.15	0.15	0.15	0.17	0.17	0.17	0.15	1.11	RI	1.58
0.2	0.2	0.2	0.2	0.2	0.19	0.1	0.19	0.190	0.163636364	0.163636364	0.163636364	0.19047619	0.19047619	0.19047619	0.20	0.87	CR	0.04
0.05	0.06	0.06	0.1	0.05	0.1	0.1	0.19	0.119	0.11	0.11	0.11	0.12	0.12	0.12	0.09	1.50		
0.2	0.24	0.25	0.2	0.2	0.19	0.16	0.1909	0.190	0.163636364	0.163636364	0.163636364	0.19047619	0.19047619	0.19047619	0.20			
0.0	0.0	0.08	0.12	0.07	0.09	0.14	0.12	3	0.14	0.127272727	0.127272727	0.127272727	0.142857143	0.142857143	0.142857143	0.12			
0.02	0.03	0.01	0.03	0.02	0.02	0.04	0.024	0.02	0.04	0.036363636	0.036363636	0.023809524	0.023809524	0.023809524	0.03			
0.0	0.0	0.0	0.0	0.01	0.0	0.02	0.0162	0.01	0.02	0.02	0.02	0.01	0.01	0.01	0.02			

Figure 30. Normalised matrix: total life-cycle cost.

SC3- Cultural restriction on usury	Compressed Stabilized Rammed Earth blocks	Clay Products- Unfired Bricks	Reclaimed/Recycled laminated Wood Flooring and Panelling	Bamboo XL laminated Split Paneled Flooring	Fly Ash Sand Lime interlocking Paving Bricks/Block	Recycled timber clad Aluminium framed window unit	Four panel hardwood door finished with Alpilignum.	Stainless Steel Entry Door.	Reprocessed Particleboard wood chipboard to BS EN 312 Type P5,	Tongue & grooved Wooddeco Multiline ceiling tiles to BS EN 636–2]	Plasterboard on 70 mm steel studs with 50 mm 12.9 kg/m³ insulation,	Tongue & Grooved Laminated Wooden column bolted to steel plate on concrete base.	Steel Column UC	Structurally insulated timber panel system with OSB/3 each side, roofing underlay reclaimed clay tiles	Structurally insulated natural slate (temperate EN 636–2) decking each side]
Compressed Stabilized Rammed Earth blocks	1.0	1.0	1.0	1.0	1.0	0.3	0.3	0.2	1.0	1.0	0.3	1.0	0.1	1.0	1.0
Clay Products—Unfired Bricks	1.0	1.0	1.0	1.0	1.0	0.3	0.3	0.2	1.0	1.0	0.3	1.0	0.1	1.0	1.0
Reclaimed/Recycled laminated Wood Flooring and Panelling	1.0	1.0	1.0	1.0	1.0	0.3	0.3	0.2	1.0	1.0	0.3	1.0	0.1	1.0	1.0
Bamboo XL laminated Split Paneled Flooring	1.0	1.0	1.0	1.0	1.0	0.3	0.3	0.2	1.0	1.0	0.3	1.0	0.1	1.0	1.0
Fly Ash Sand Lime interlocking Paving Bricks/Block	1.0	1.0	1.0	1.0	1.0	0.3	0.3	0.2	1.0	1.0	0.3	1.0	0.1	1.0	1.0

Continued

Recycled timber clad Aluminium framed window unit	3.0	3.0	3.0	3.0	3.0	1.0	1.0	0.3	3.0	3.0	1.0	3.0	0.2	3.0	3.0
Four panel hardwood door finished with Alpilignum.	3.0	3.0	3.0	3.0	3.0	1.0	1.0	0.3	3.0	3.0	1.0	3.0	0.2	3.0	3.0
Stainless Steel Entry Door.	5.0	5.0	5.0	5.0	5.0	3.0	3.0	1.0	5.0	5.0	3.0	5.0	0.3	5.0	5.0
Reprocessed Particleboard wood chipboard to BS EN 312 Type P5,	1.0	1.0	1.0	1.0	1.0	0.3	0.3	0.2	1.0	1.0	0.3	1.0	0.1	1.0	1.0
Tongue & grooved Wooddeco Multiline ceiling tiles to BS EN 636–2]	1.0	1.0	1.0	1.0	1.0	0.3	0.3	0.2	1.0	1.0	0.3	1.0	0.1	1.0	1.0
Plasterboard on 70 mm steel studs with 50 mm 12.9 kg/m³ insulation,	3.0	3.0	3.0	3.0	3.0	1.0	1.0	0.3	3.0	3.0	1.0	3.00	0.20	3.00	3.00
Tongue & Grooved Laminated Wooden column bolted to steel plate on concrete base.	1.0	1.0	1.0	1.0	1.0	0.3	0.3	0.2	1.0	1.0	0.3	1.0	0.1	1.0	1.0
Steel Column UC	7.0	7.0	7.0	7.0	7.0	5.0	5.0	3.0	7.0	7.0	5.0	7.0	1.0	7.0	7.0
Structurally insulated timber panel system with OSB/3 each side, roofing underlay reclaimed clay tiles	1.0	1.0	1.0	1.0	1.0	0.3	0.3	0.2	1.0	1.0	0.3	1.0	0.1	1.0	1.0
Structurally insulated natural slate (temperate EN 636-2) decking each side]	1.0	1.0	1.0	1.0	1.0	0.3	0.3	0.2	1.0	1.0	0.3	1.0	0.1	1.0	1.0
Total	31.0	31.0	31.0	31.0	31.0	14.3	14.3	7.0	31.0	31.0	14.3	31.0	3.4	31.0	31.0

Figure 31. Pair-wise matrix: cultural restriction on usury.

Compressed Stabilized Rammed Earth blocks	Clay Products- Unfired Bricks	Reclaimed/Recycled laminated Wood Flooring and Panelling	Bamboo XL laminated Split Paneled Flooring	Fly Ash Sand Lime interlocking Paving Bricks/Block	Recycled timber clad Aluminium framed window unit	Four panel hardwood door finished with Alpilignum.	Stainless Steel Entry Door.	Reprocessed Particleboard wood chipboard to BS EN 312 Type P5,	Tongue & grooved Wooddeco Multiline ceiling tiles to BS EN 636–2]	Plasterboard on 70 mm steel studs with 50 mm 12.9 kg/m³ insulation,	Tongue & Grooved Laminated Wooden column bolted to steel plate on concrete base.	Steel Column UC	Structurally insulated timber panel system with OSB/3 each side, roofing underlay reclaimed clay tiles	Structurally insulated natural slate (temperate EN 636-2) decking each side]	Average	Lambda Max	CI	0.02
0.0	0.0	0.0	0.0	0.03	0.0	0.02	0.02	0.03	0.03	0.02	0.03	0.04	0.03	0.03	0.03	0.96	RI	1.58
0.0	0.0	0.0	0.0	0.03	0.0	0.02	0.02	0.03	0.03	0.02	0.03	0.04	0.03	0.03	0.03	0.96	CR	0.01
0.0	0.0	0.0	0.0	0.03	0.0	0.02	0.02	0.03	0.03	0.02	0.03	0.04	0.03	0.03	0.03	0.96		
0.0	0.0	0.0	0.0	0.03	0.0	0.02	0.02	0.03	0.03	0.02	0.03	0.04	0.03	0.03	0.03	0.96		
0.0	0.0	0.0	0.0	0.03	0.0	0.02	0.02	0.03	0.03	0.02	0.03	0.04	0.03	0.03	0.03	0.96		
0.1	0.1	0.1	0.1	0.10	0.1	0.07	0.04	0.09	0.10	0.07	0.10	0.06	0.10	0.10	0.09	1.23		
0.1	0.1	0.1	0.1	0.10	0.1	0.07	0.04	0.09	0.10	0.07	0.10	0.06	0.10	0.10	0.09	1.23		
0.2	0.2	0.2	0.2	0.16	0.2	0.21	0.14	0.16	0.16	0.21	0.16	0.10	0.16	0.16	0.17	1.16		
0.0	0.0	0.0	0.0	0.03	0.0	0.02	0.02	0.03	0.03	0.02	0.03	0.04	0.03	0.03	0.03	0.96		
0.0	0.0	0.0	0.0	0.0	0.0	0.0	0.02	0.03	0.03	0.02	0.03	0.04	0.03	0.03	0.03	0.96		
0.10	0.10	0.10	0.1	0.10	0.1	0.1	0.04	0.09	0.10	0.07	0.10	0.06	0.10	0.10	0.09	1.23		
0.0	0.0	0.0	0.0	0.0	0.0	0.0	0.028	0.03	0.03	0.02	0.03	0.04	0.03	0.03	0.03	0.96		

Continued

0.2	0.2	0.2	0.2	0.2	0.3	0.3	0.42	0.22	0.23	0.35	0.23	0.30	0.23	0.23	0.27	0.90
0.0	0.0	0.0	0.0	0.0	0.0	0.0	0.02	0.03	0.03	0.02	0.03	0.04	0.03	0.03	0.03	0.96
0.0	0.0	0.0	0.0	0.03	0.0	0.02	0.02	0.03	0.03	0.02	0.03	0.04	0.03	0.03	0.03	0.96
1.00	1.00	1.00	1.00	1.00	1.00	1.00	1.00	1.00	1.00	1.00	1.00	1.00	1.00	1.00	1.00	15.3

Figure 32. Normalised matrix: cultural restriction on usury.

T2-Ease to remove/reaffix/replace	Compressed Stabilized Rammed Earth blocks	Clay Products- Unfired Bricks	Reclaimed/Recycled laminated Wood Flooring and Panelling	Bamboo XL laminated Split Paneled Flooring	Fly Ash Sand Lime interlocking Paving Bricks/Block	Recycled timber clad Aluminium framed window unit	Four panel hardwood door finished with Alpilignum.	Stainless Steel Entry Door.	Reprocessed Particleboard wood chipboard to BS EN 312 Type P5,	Tongue & grooved Wooddeco Multiline ceiling tiles to BS EN 636–2]	Plasterboard on 70 mm steel studs with 50 mm 12.9 kg/m³ insulation,	Tongue & Grooved Laminated Wooden column bolted to steel plate on concrete base.	Steel Column UC	Structurally insulated timber panel system with OSB/3 each side, roofing underlay reclaimed clay tiles	Structurally insulated natural slate (temperate EN 636-2) decking each side]
Compressed Stabilized Rammed Earth blocks	1.0	0.3	0.2	0.2	0.3	0.2	0.2	0.3	0.2	0.2	0.3	0.20	0.33	0.20	0.20
Clay Products—Unfired Bricks	3.0	1.0	0.3	0.3	0.5	0.3	0.3	1.0	0.3	0.3	1.0	0.3	1.0	0.3	0.3
Reclaimed/Recycled laminated Wood Flooring and Panelling	5.0	3.0	1.0	1.0	2.0	1.0	1.0	3.0	1.0	1.0	3.0	1.0	3.0	1.0	1.0
Bamboo XL laminated Split Paneled Flooring	5.0	3.0	1.0	1.0	2.0	1.0	1.0	3.0	1.0	1.0	3.0	1.0	3.0	1.0	1.0
Fly Ash Sand Lime interlocking Paving Bricks/Block	4.0	2.0	0.5	0.5	1.0	0.5	0.5	2.0	0.5	0.5	2.0	0.5	2.0	0.5	0.5
Recycled timber clad Aluminium framed window unit	5.0	3.0	1.0	1.0	2.0	1.0	1.0	3.0	1.0	1.0	3.0	1.0	3.0	1.0	1.0
Four panel hardwood door finished with Alpilignum.	5.0	3.0	1.0	1.0	2.0	1.0	1.0	3.0	1.0	1.0	3.0	1.0	3.0	1.0	1.0
Stainless Steel Entry Door.	3.0	1.0	0.3	0.3	0.5	0.3	0.3	1.0	0.3	0.3	1.0	0.3	1.0	0.3	0.3
Reprocessed Particleboard wood chipboard to BS EN 312 Type P5,	5.0	3.0	1.0	1.0	2.0	1.0	1.0	3.0	1.0	1.0	3.0	1.0	3.0	1.0	1.0
Tongue & grooved Wooddeco Multiline ceiling tiles to BS EN 636–2]	5.0	3.0	1.0	1.0	2.0	1.0	1.0	3.0	1.0	1.0	3.0	1.0	3.0	1.0	1.0
Plasterboard on 70 mm steel studs with 50 mm 12.9kg/m³ insulation,	3.0	1.0	0.3	0.3	0.5	0.3	0.3	1.0	0.3	0.3	1.0	0.3	1.0	0.3	0.3
Tongue & Grooved Laminated Wooden column bolted to steel plate on concrete base.	5.0	3.0	1.0	1.0	2.0	1.0	1.0	3.0	1.0	1.0	3.0	1.0	3.0	1.0	1.0
Steel Column UC	3.0	1.0	0.3	0.3	0.5	0.3	0.3	1.0	0.3	0.3	1.0	0.3	1.0	0.3	0.3
Structurally insulated timber panel system with OSB/3 each side, roofing underlay reclaimed clay tiles	5.0	3.0	1.0	1.0	2.0	1.0	1.0	3.0	1.0	1.0	3.0	1.0	3.0	1.0	1.0
Structurally insulated natural slate (temperate EN 636-2) decking each side]	5.0	3.0	1.0	1.0	2.0	1.0	1.0	3.0	1.0	1.0	3.0	1.00	3.00	1.00	1.0
Total	62.0	33.3	11.0	11.0	21.3	11.0	11.0	33.3	11.0	11.0	33.3	11.0	33.3	11.0	11.0

Figure 33. Pair-wise matrix: ease to remove/affix/replace.

Compressed Stabilized Rammed Earth blocks	Clay Products- Unfired Bricks	Reclaimed/Recycled laminated Wood Flooring and Panelling	Bamboo XL laminated Split Paneled Flooring	Fly Ash Sand Lime interlocking Paving Bricks/Block	Recycled timber clad Aluminium framed window unit	Four panel hardwood door finished with Alpilignum.	Stainless Steel Entry Door.	Reprocessed Particleboard wood chipboard to BS EN 312 Type P5,	Tongue & grooved Wooddeco Multiline ceiling tiles to BS EN 636–2]	Plasterboard on 70 mm steel studs with 50 mm 12.9 kg/m³ insulation,	Tongue & Grooved Laminated Wooden column bolted to steel plate on concrete base.	Steel Column UC	Structurally insulated timber panel system with OSB/3 each side, roofing underlay reclaimed clay tiles	Structurally insulated natural slate (temperate EN 636–2) decking each side]	Average	Lambda Max	CI 0.01
0.02	0.01	0.02	0.0	0.01	0.0	0.0	0.01	0.01	0.02	0.01	0.02	0.01	0.02	0.02	0.02	0.95	RI
0.0	0.0	0.0	0.0	0.0	0.0	0.0	0.03	0.03	0.03	0.03	0.03	0.03	0.03	0.03	0.03	1.03	CR
0.1	0.1	0.1	0.1	0.1	0.1	0.1	0.09	0.09	0.09	0.09	0.09	0.09	0.09	0.09	0.09	0.99	
0.1	0.1	0.1	0.1	0.1	0.1	0.1	0.09	0.09	0.09	0.09	0.09	0.09	0.09	0.09	0.09	0.99	
0.1	0.1	0.0	0.0	0.05	0.0	0.05	0.06	0.04	0.05	0.06	0.05	0.06	0.05	0.05	0.05	1.08	
0.1	0.1	0.1	0.1	0.09	0.1	0.09	0.09	0.09	0.09	0.09	0.09	0.09	0.09	0.09	0.09	0.99	
0.1	0.1	0.1	0.1	0.09	0.1	0.09	0.09	0.09	0.09	0.09	0.09	0.09	0.09	0.09	0.09	0.99	
0.0	0.0	0.0	0.0	0.02	0.0	0.03	0.03	0.03	0.03	0.03	0.03	0.03	0.03	0.03	0.03	1.03	
0.1	0.1	0.1	0.1	0.09	0.1	0.09	0.09	0.09	0.09	0.09	0.09	0.09	0.09	0.09	0.09	0.99	
0.1	0.1	0.1	0.1	0.09	0.1	0.09	0.09	0.09	0.09	0.09	0.09	0.09	0.09	0.09	0.09	0.99	
0.0	0.0	0.0	0.0	0.02	0.0	0.03	0.03	0.03	0.03	0.03	0.03	0.03	0.03	0.03	0.03	1.03	
0.1	0.1	0.1	0.1	0.09	0.1	0.09	0.09	0.09	0.09	0.09	0.09	0.09	0.09	0.09	0.09	0.99	
0.0	0.0	0.0	0.0	0.02	0.0	0.03	0.03	0.03	0.03	0.03	0.03	0.03	0.03	0.03	0.03	1.03	
0.1	0.1	0.1	0.1	0.1	0.1	0.1	0.09	0.09	0.09	0.09	0.09	0.09	0.09	0.09	0.09	0.99	
0.08	0.09	0.09	0.1	0.09	0.1	0.1	0.09	0.09	0.09	0.09	0.09	0.09	0.09	0.09	0.09	0.99	
1.00	1.00	1.00	1.00	1.00	1.00	1.00	1.00	1.00	1.00	1.00	1.00	1.00	1.00	1.00	1.00	15.1	

Figure 34. Normalised matrix: ease to remove/affix/replace.

SN5- Acoustics Performance	Compressed Stabilized Rammed Earth blocks	Clay Products- Unfired Bricks	Reclaimed/Recycled laminated Wood Flooring and Panelling	Bamboo XL laminated Split Paneled Flooring	Fly Ash Sand Lime interlocking Paving Bricks/Block	Recycled timber clad Aluminium framed window unit	Four panel hardwood door finished with Alpilignum.	Stainless Steel Entry Door.	Reprocessed Particleboard wood chipboard to BS EN 312 Type P5,	Tongue & grooved Wooddeco Multiline ceiling tiles to BS EN 636–2]	Plasterboard on 70 mm steel studs with 50 mm 12.9 kg/m³ insulation,	Tongue & Grooved Laminated Wooden column bolted to steel plate on concrete base.	Steel Column UC	Structurally insulated timber panel system with OSB/3 each side, roofing underlay reclaimed clay tiles	Structurally insulated natural slate (temperate EN 636–2) decking each side]
Compressed Stabilized Rammed Earth blocks	1.0	0.2	0.3	0.2	0.2	0.3	0.2	1.0	0.3	0.3	0.3	0.3	0.3	0.3	1.0
Clay Products—Unfired Bricks	5.0	1.0	2.0	1.0	1.0	3.0	1.0	5.0	2.0	2.0	2.0	2.0	2.0	2.0	5.0

Continued

Reclaimed/Recycled laminated Wood Flooring and Panelling	4.0	0.5	1.0	0.5	0.5	2.0	0.5	4.0	1.0	1.0	1.0	1.0	1.0	1.0	4.0
Bamboo XL laminated Split Paneled Flooring	5.0	1.0	2.0	1.0	1.0	3.0	1.0	5.0	2.0	2.0	2.0	2.0	2.0	2.0	5.0
Fly Ash Sand Lime interlocking Paving Bricks/Block	5.0	1.0	2.0	1.0	1.0	3.0	1.0	5.0	2.0	2.0	2.0	2.0	2.0	2.0	5.0
Recycled timber clad Aluminium framed window unit	3.0	0.3	0.5	0.3	0.3	1.0	0.3	3.0	0.5	0.5	0.5	0.5	0.5	0.5	3.0
Four panel hardwood door finished with Alpilignum.	5.0	1.0	2.0	1.0	1.0	3.0	1.0	5.0	2.0	2.0	2.0	2.0	2.0	2.0	5.0
Stainless Steel Entry Door.	1.0	0.2	0.3	0.2	0.2	0.3	0.2	1.0	0.3	0.3	0.3	0.3	0.3	0.3	1.0
Reprocessed Particleboard wood chipboard to BS EN 312 Type P5,	4	0.5	1	0.5	0.5	2	0.5	4	1	1	1	1	1	1	4
Tongue & grooved Wooddeco Multiline ceiling tiles to BS EN 636–2]	4.0	0.5	1.0	0.5	0.5	2.0	0.5	4.0	1.0	1.0	1.0	1.0	1.0	1.0	4.0
Plasterboard on 70 mm steel studs with 50 mm 12.9 kg/m³ insulation,	4	0.5	1	0.5	0.5	2	0.5	4	1	1	1	1	1	1	4
Tongue & Grooved Laminated Wooden column bolted to steel plate on concrete base.	4	0.5	1	0.5	0.5	2	0.5	4	1	1	1	1	1	1	4
Steel Column UC	4.00	0.50	1.00	0.50	0.50	2.00	0.50	4.00	1.00	1.00	1.00	1.00	1.00	1.00	4.00
Structurally insulated timber panel system with OSB/3 each side, roofing underlay reclaimed clay tiles	4.0	0.5	1.0	0.5	0.5	2.0	0.5	4.0	1.0	1.0	1.0	1.0	1.0	1.0	4.0
Structurally insulated natural slate (temperate EN 636-2) decking each side]	1.0	0.2	0.3	0.2	0.2	0.3	0.2	1.0	0.3	0.3	0.3	0.3	0.3	0.3	1.0
Total	54.0	8.4	16.3	8.4	8.4	28.0	8.4	54.0	16.3	16.3	16.3	16.3	16.3	16.3	54.0

Figure 35. Pair-wise matrix: acoustics performance.

Compressed Stabilized Rammed Earth blocks	Clay Products- Unfired Bricks	Reclaimed/Recycled laminated Wood Flooring and Panelling	Bamboo XL laminated Split Paneled Flooring	Fly Ash Sand Lime interlocking Paving Bricks/Block	Recycled timber clad Aluminium framed window unit	Four panel hardwood door finished with Alpilignum.	Stainless Steel Entry Door.	Reprocessed Particleboard wood chipboard to BS EN 312 Type P5,	Tongue & grooved Wooddeco Multiline ceiling tiles to BS EN 636–2]	Plasterboard on 70 mm steel studs with 50 mm 12.9 kg/m³ insulation,	Tongue & Grooved Laminated Wooden column bolted to steel plate on concrete base.	Steel Column UC	Structurally insulated timber panel system with OSB/3 each side, roofing underlay reclaimed clay tiles	Structurally insulated natural slate (temperate EN 636-2) decking each side]	Average	Lambda Max	CI	0.01
0.0	0.0	0.0	0.0	0.0	0.0	0.0	0.01	0.01	0.02	0.02	0.02	0.02	0.02	0.02	0.02	0.97	RI	1.58
0.1	0.1	0.1	0.1	0.1	0.1	0.1	0.09	0.12	0.12	0.12	0.12	0.12	0.12	0.09	0.11	0.97	CR	0.01
0.1	0.1	0.1	0.1	0.1	0.1	0.1	0.07	0.061	0.06	0.06	0.06	0.06	0.06	0.07	0.06	1.04		
0.1	0.1	0.1	0.1	0.1	0.1	0.1	0.09	0.123	0.12	0.12	0.12	0.12	0.12	0.09	0.11	0.97		
0.1	0.1	0.1	0.1	0.1	0.1	0.1	0.09	0.123	0.12	0.12	0.12	0.12	0.12	0.09	0.11	0.97		
0.1	0.0	0.0	0.0	0.0	0.0	0.0	0.05	0.030	0.03	0.03	0.03	0.03	0.03	0.06	0.04	1.07		
0.1	0.1	0.1	0.1	0.1	0.1	0.1	0.09	0.123	0.12	0.12	0.12	0.12	0.12	0.09	0.11	0.97		
0.0	0.0	0.0	0.0	0.0	0.0	0.0	0.018	0.01	0.02	0.02	0.02	0.02	0.02	0.02	0.02	0.97		

Continued

0.074	0.059	0.06	0.059	0.059	0.071	0.059	0.074	0.061	0.061	0.061	0.061538462	0.62	0.061538462	0.074074074	0.06	1.04
0.1	0.1	0.1	0.1	0.1	0.1	0.1	0.074	0.061	0.06	0.06	0.06	0.06	0.06	0.07	0.06	1.04
0.07	0.059	0.06	0.059	0.059	0.071	0.059	0.074	0.061	0.061	0.061	0.061538462	0462	0.061538462	0.074074074	0.06	1.04
0.07	0.059	0.06	0.059	0.059	0.071	0.059	0.074	0.061	0.061	0.061	0.061538462	0.0	0.061538462	0.074074074	0.06	1.04
0.07	0.06	0.06	0.06	0.06	0.07	0.06	0.074	0.06	0.061	0.061	0.061538462	62	0.061538462	0.074074074	0.06	1.04
0.1	0.1	0.1	0.1	0.1	0.1	0.1	0.074	0.061	0.06	0.06	0.06	0.06	0.06	0.07	0.06	1.04
0.0	0.0	0.0	0.0	0.0	0.0	0.0	0.018518519	0.015384615	0.02	0.02	0.02	0.02	0.02	0.02	0.02	0.97
1.00	1.00	1.00	1.00	1.00	1.00	1.00	1.00	1.00	1.00	1.00	1.00	1.00	1.00	1.00	1.00	15.2

Figure 36. Normalised matrix: acoustics performance.

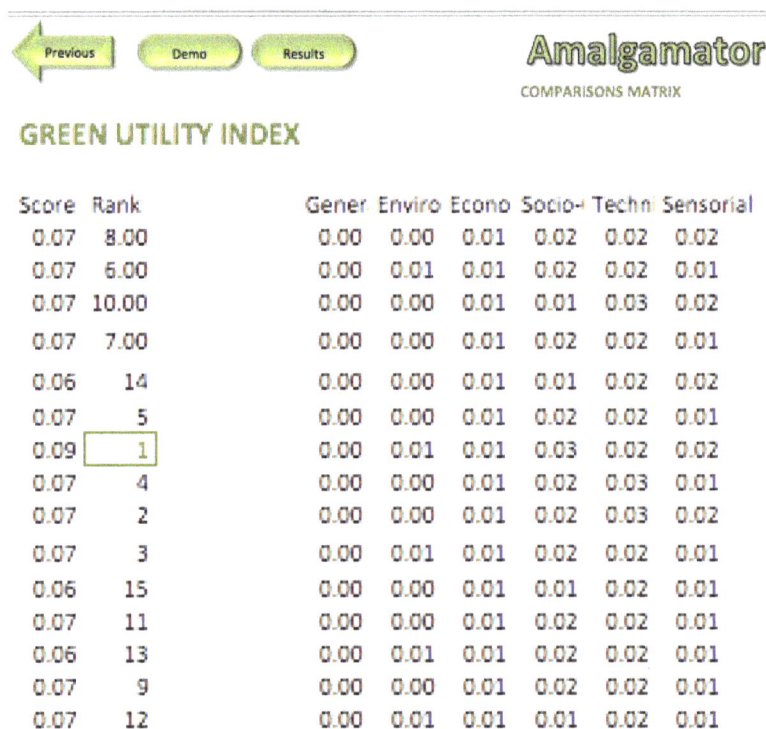

Figure 37. Green utility indices of the selected materials.

6. Potential Benefits of the MSDSS Model

The following are the benefits expected from the application of the MSDSS Model. However the model developed for this research differs from that of the previous works in the following ways:

- The main point of difference from the off-the-shelf assessment tools is that they only trade-off numerical values based on the single-attributes. These single-attribute claims ignore the possibility of what other variables can yield. MSDSS supports trade-off with and without tangible variables, such as a client's preference, environmental statutory compliance, and cultural restriction on usury. This feature is important as decision making in reality engages with solid, verbal and subjective elements.

- In terms of cost, it provides an opportunity for designers to be able to advise their clients as to what the probable financial estimate of the project may be. This helps clients to decide how much they are prepared to spend on different variables of construction.

- A separate set of contextual considerations was included as a heuristics base to facilitate site-specific feasibility and appropriateness testing of each material choice. Boundaries of sustainability inform of knowledge base rules as contained in the MSDSS model could help reduce bias that is often associated with the material selection process.

- Available material assessment tools are particularity ill-adapted for the early stages of the design process and are generally labour intensive. The MSDSS model consists of a resource for relatively small

Results

Fly Ash Cement concrete Floor slab

- General/Site
- Environment/Health
- Economic/Cost
- Socio-Cultural
- Technical
- Sensorial

Candidate Materials	Green Utility Index	Material Category
Fly Ash Cement concrete Floor slab	8.6%	Conventional Material
Fly Ash Sand Lime interlocking Paving Bricks/Block	7.3%	Locally-Sourced Material
Four panel hardwood door finished with Aipilignum.	7.3%	Locally-Sourced Material
Recycled Porocom Pebble Stone Pavers	7.2%	Recycled Material
Bamboo XL laminated Split Paneled Flooring	7.2%	Locally-Sourced Material
Compressed Stabilized Rammed Earth blocks	7.1%	Locally-Sourced Material
Clay Products- Unfired Bricks	6.8%	Locally-Sourced Material
Recycled crushed concrete block	6.7%	Recycled Material
Structurally insulated natural slate (temperate EN 636-2) decking each side|	6.7%	Locally-Sourced Material
Insulated Concrete Form block (ICF)/cementbound recycled wood chip	6.7%	Recycled Material
Tongue & grooved Wooddeco Multiline ceiling tiles to BS EN 636-2|	6.6%	Locally-Sourced Material
Roofing underlay concrete interlocking tiles	6.5%	Conventional Material
Tongue & Grooved Laminated Wooden column bolted to steel plate on concrete base	6.5%	Locally-Sourced Material
Reclaimed/Recycled laminated Wood Flooring and Panelling	6.2%	Recycled Material
Reprocessed Particleboard wood chipboard to BS EN 312 Type P5,	5.8%	Recycled Material

Material Properties — Fly Ash Cement concrete Floor slab

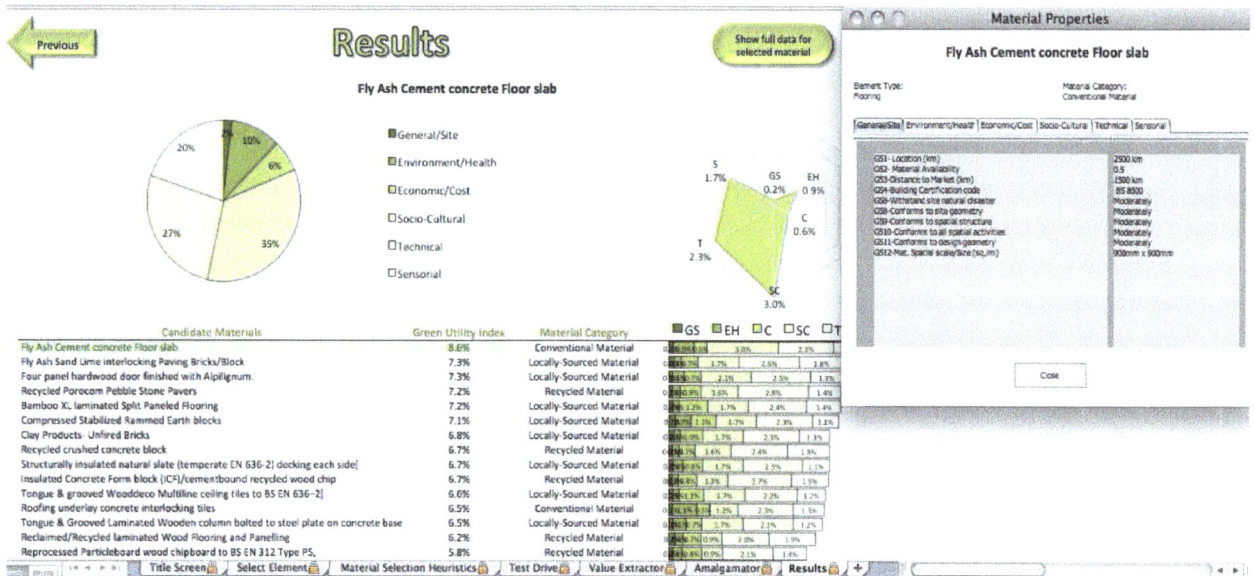

Figure 38. Corresponding indices of the ranked materials.

information input to produce quick and fairly accurate or approximate output of results with little or no training on the part of experienced users. This means that users that may require little training are inexperienced users but not as extensive as obtainable in previous tools.

- There are still significant numbers of smaller firms who cannot afford most material assessment tools because they are extremely expensive. This tool is more or less open source software recommended to provide solution to this challenge.
- Context is a critical consideration for all project decision-making, since even projects located on neighbouring sites will have different end users, and different specific site characteristics. This tool could be applied to other regions with minimal or no changes, and therefore has the ability to adapt to any situation, or change in design according to users' needs or different material alternatives.
- Unlike in the previous models, this tool contains tutorials and help menu as well as video guidance on how to use the software. This provides adequate help to beginners or inexperienced designers.
- For the visual aspect, the MSDSS model has the ability to produce a picture representative of data input rather than abstract. It is able to transfer data from it to other software, applicable to building material selection, and present the properties of each material in a successive window.
- User weightings have been included in the selection methodology to supplement, and not supplant human judgment in the decision-making process. By incorporating user weightings into the selection process, the methodology gains greater acceptability to the user who supplies the weightings.

- Materials change in their innovation, composition, price and availability and most tools find it challenging to update information relating to products. In this MSDSS model, the materials and the corresponding performance of the selected products is updated through a link to the manufacturers web page on the internet, and the users may access more information regarding the selected material or technology through internet from the supplier's web pages.
- The system has been designed to produce an artistic output, accurate, detailed representation and close to reality as much as it can be, without attempt to conceal any feature whether attractive or not;
- Provision of only a limited set of operations or criteria restricts the techniques and solutions that can be applied and consequently restricts the decision-making process. On the other hand, the inclusion of many objectives and the permitting of user specification of input data, system parameters and models, generally increases system flexibility and increases decision support freedom;
- In most tools, AHP technique at the pare-wise comparison stage, tend to be quite cumbersome and often takes a lot of time to maintain the consistency of the response. To eliminate this challenge MSDSS automatically debugs the system at every stage of the evaluation and selection process.
- The system has been thoroughly debugged to be less error prone, so that practitioners can integrate the decisions made by the tools more smoothly into practice, and that it takes less than few seconds to respond to users inputs;
- Responses/feedback from system programmers and

accredited green building experts have also been included in the study to prove the ease of use, applicability and usability of the MSDSS model (see appendix A). As a result, some features have been adjusted based on expert feedbacks to support more reliable and expedient, timelier feedback to different design alternatives or changes.

Reflective Summary

This paper discussed the process of developing a decision-support system to support choices in low-cost green building materials. The research presented in this paper acknowledged the lack of a reliable database model that decision makers can readily use to aid informed decision-making when selecting low-cost green materials for low-cost green residential housing development. The findings from the reviewed literature and the results of the surveyed questionnaire further underscored the need for improving understanding of relevant data associated with the use of such building materials and components, with the goal to change and positively influence the current mental models, attitudes and priorities of multiple stakeholders involved in the production of the built environment, so as to encourage their wider-scale use in mainstream housing.

Based on the data obtained from selected expert builder/developer companies, a prototype MSDSS model was developed to aid designers in making informed decisions regarding their choice of materials for low-cost green residential housing projects. This model was consolidated in to an excel-based decision tool that allows designers to select low-cost green building products from a range of possibilities, and view the resulting impacts and difference in the cost, durability and performance of a range of alternatives. An analysis using the Analytical Hierarchy Process (AHP), based on the results of the participants was performed to show how optimal choices could change with changing user weightings and variables. The participants gained views from participating in the evaluation exercise for a real-life project, including the difficulties in choosing preference scores.

This study thus, indicates that perhaps the development of a DSS model associated with the impacts of low-cost green building materials is useful in that it gives designers a new approach of going through the process of value elicitation, which allows them to explicitly and transparently test the impacts of their elicited values. Providing a visual representation, allowing designers or specifiers to compare multiple alternatives across multiple criteria, was a particularly useful aspect of this study.

7. Conclusions

This report has demonstrated how a DSS model can be used to support multi-stakeholder involvement in the selection of low-cost green construction materials in ways that enable building energy performance and life-cycle cost to be considered at the early stage of residential housing design. The study further reinforced the significance in taking a multi-attribute approach to assessing a building product's sustainable performance. To achieve this goal, the AHP model of decision-making [57-60] was adopted to deal with the ambiguities involved in the assessment of material alternatives and relative importance weightings of multiple factors, given its ability to solve multi-criteria decision-making (MCDM) between finite alternatives.

To prove the validity of the model and the feasibility of the proposed selection methodology, a real-life but hypothetical application scenario was used to further illustrate the application of the MSDSS model in selecting the most appropriate floor material for a single 5-bedroom residential housing project located in the Sutton County of London. The results demonstrated the capabilities of the system, and exposed the way in which the system transparently demonstrates the implications of each step of the analysis. It also proved the practicality of using the MSDSS model, as it combines multiple factors into a single performance value that is easily interpreted.

Since the purpose of this research study was to develop an innovative concept to demonstrate a step-by-step methodology for selecting low-cost green materials with reasonable accuracy and in real time, as opposed to developing a fully-equipped commercial software, macro-in-excel database management technique was used in the back-end of the system to integrate the large volumes of data obtained from multiple sources. Excel was adopted as the database management system since it has the capabilities to perform all necessary calculations and is common enough that most people are familiar with it.

The process followed to develop the prototype MSDSS model in this research demonstrates that, depending on the domain and scope of the problem at hand, a DSS can be built fairly quickly and can be used effectively to help designers quantify how they compare materials that are yet to be certified under the standard specifications and codes of practice, and that which are already permitted under existing codes.

However further work is required to fully validate the MSDSS and the methodology presented. To do so, this research intends to run further case studies ideally using "live" building design projects, by comparing the outputs from the algorithms of the MSDSS system to monitored data from the completed case study building, in order to review the potential savings of the new materials or components proposed by the MSDSS model.

7.1. Contributions to Research and Industry

Insights identified from addressing the research objectives in Section 3 represent part of the original contribution to knowledge made by this study. The following are itemised as key contributions of the study to research and practice:

- The contribution of this research includes the consideration of a holistic approach to low-cost green building product selection based on socio-cultural, technical, emotive, site, cost and environmental performance. Pre-design estimators and pre-construction managers could improve their estimating and product selection practices using the proposed MSDSS tool.
- Material suppliers can also benefit from this approach, as they can use it to enhance their pricing strategies, marketing plans, and overall product competitiveness.
- Decision problems about a product's choice are usually unstructured and ill-defined. By suggesting an alternative means of integrating the available resources associated with the informed selection of low-cost green building materials, it is hoped that the model will help decision makers to further refine their material selection criteria thus, encourage effective decision-making.
- The material selection process is characterized by competitive objectives, involving multiple stakeholders and key actors, dynamic and uncertain procedures and limited timeframes to make significant decisions. The decision makers within this domain: the designers, specifiers and other stakeholders are often confronted with conflicting subjective preferences and fragmented expertise; hence resulting in decision-making failures. The capacity of the system to compare materials using multiple factors with user-specified weightings, will therefore, encourage decision-makers to explicitly consider the effects of their previously-implicit judgments on the outcome of the project, and thus make choices that are timely, and result in more sustainable residential housing project design and implementation.
- The ability to quickly quantify and qualify the suitability outcomes of alternative materials may encourage greater industry acceptance of innovative technology for materials that are yet to be certified under the standard specifications and codes of practice.
- The overall approach used here could be tested in other contexts to determine its generalizability and applicability. In other words, the system could be extended to select materials for commercial development or for any other purpose.
- The material selection factors identified in the prototype model of the MSDSS, provides a unique insight into sustainability and environmental design informa-

tion requirements for low-cost green housing.
- The adopted research methodology (see **Table 1**) employed to address the research objectives in Section 3 represents part of the original contribution to knowledge made by this study.
- The number of academic publications on the impacts of low-cost green materials was found to be low; hence makes a crucial contribution.
- In the short term, the model could be used in the housing sector as a catalogue of materials to support decision-making in low-cost green housing designs.
- As low-cost green building materials and components become well understood by design and building professionals, there is a likelihood of reducing over-dependency on conventional construction materials in the housing industry.
- The outcome of this study could aid top executives within the housing sector to consider low-cost green materials as part of existing regulatory frameworks and building codes of the Construction Standards Institute (CSI) in capital projects. By doing so, such an approach may create a potential market for local manufacturing and processing of such materials.

7.2. Setbacks, Challenges and Probable Solutions

There were few possible limitations that this research faced during the cause of the study. The limitations are hereby listed for future consideration.

- The process of developing the selection methodology was faced with critical issues that led to several changes in the research methodology and its objectives so many times, in order to achieve the aim of this research.
- Citing prior research studies formed the basis of the literature review and helped lay the foundation for understanding the research problem investigated in this study. However, there were reservations regarding the currency and scope of the research topic, as there was no compelling evidence of prior research on the topic. As literature on DSS for low-cost green housing design is still relatively low, the study therefore had to rely on the most current reports, interviews, and observations from the different and various organisations, and building professionals for its information.
- It remains true that sample sizes that are too small cannot adequately support claims of having achieved valid conclusions and sample sizes that are too large do not permit the deep, naturalistic, and inductive analysis that defines qualitative inquiry [47]. Yin [47] noted that determining adequate sample size in qualitative research is ultimately a matter of judgment and experience in evaluating the quality. Hair *et al.* [61] warned that it is important to consider not only the

statistical significance, but also the quality and practical significance of the results for managerial applications, when analysing data. They noted that unequal or uneven sample sizes amongst different professional groups could also bias or influence the results as getting equal sample sizes from different groups of respondents was unrealistic and demanding. To address this issue the study adopted a sampling strategy using the stratified random sampling approach where each group of the sample population had reasonable number of randomly selected participants, which helped to achieve sampling equivalence between the researcher and professionals of the various building professions both in higher institutions and practicing building design and housing construction firms.

- Giving that most respondents were practicing professionals, getting a list of the sample population for the study was very discouraging. Having access to people, and organizations, was otherwise limited, giving the time differences and tight-scheduled activities. However the use of progressive approach of reminding the subjects using any available means either through e-mails, LinkedIn, Facebook, Twitter or through phone calls helped to address this problem.

- Very few of the participants had little exposure to AHP quantitative-based decision-making process. Though they found the process a bit daunting, they were somewhat comfortable with the idea of ranking preferences, as they were used to considering the choice for alternatives based on unquantified methods, but without assigning personal values to criteria. Prior help manual sent to participants before embarking on expert evaluation survey helped to reduce the complexities associated with the MCDM technique adopted.

7.3. Potential Areas for Further Studies

Several areas were identified as potential areas for further research as itemised below:

- Although not demonstrated in this system but it is also possible that potential researchers can redesign or customize the database to best fit the needs of any particular region or could be extended to select materials for commercial development;

- While the findings of this research focused specifically on a subset of design and building professionals involved with public residential housing sector projects, the overall approach used here could be tested in other contexts to determine its generalizability and applicability.

8. Acknowledgements

This work was made possible through private funding,

and was partially supported by a discount from the Journal of Building Construction and Planning Research Doctoral Research Student Discount Scheme program.

REFERENCES

[1] IEA (International Energy Agency), "Energy Efficiency Requirements in Building Codes, Energy Efficiency Policies for New Buildings," OECD/IEA, Paris, 2008.

[2] IEA (International Energy Agency), "IEA Net Zero Energy," Montreal, 2009.

[3] World Bank, "Nigeria: State Building, Sustaining Growth, and Reducing Poverty. A Country Economic Report," Report 29551-NG, Poverty Reduction and Economic Management Sector Unit, West Africa Region, Washington DC, 2010.

[4] UN-HABITAT, "Global Campaign on Urban Governance," Oxford University Press, New York, 2011. http://www.unhabitat.org.

[5] United Nations Development Plan (UNDP), "African Economic Outlook 2011: Africa and Its Emerging Partners," African Development Bank, OECD, UNDP and UNECA, 2011.

[6] United States Department of Energy (USDOE), "Energy Efficiency and Renewable Energy," Federal Energy Management Program, 2010, pp. 1-34.

[7] United States Department of Energy, "About the Weatherization Assistance Program," Washington DC, 2010. http://www1.eere.energy. gov/wip/wap.html

[8] J. Kennedy, "Building without Borders: Sustainable Construction for the Global Village," New Society Publishers, Gabriola, 2004.

[9] M. Shuman, "The Small-Mart Revolution: How Local Businesses Are Beating the Global Competition," Berrett-Koehler Publishers, San Francisco, 2008.

[10] Y. Oruwari, M. Jev and P. Owei, "Acquisition of Technological Capability in Africa: A Case Study of Indigenous Building Materials Firms in Nigeria," ATPS Working Paper Series No. 33, African Technology Policy Studies Network, Nairobi, 2002.

[11] K. K. Ashraf, "This Is Not a Building! Hand-Making a School in a Bangladeshi Village," *Architectural Design*, Vol. 77, No. 6, 2007, pp. 114-117.

[12] C. C. Zhou, G. F. Yin and X. B. Hu, "Multi-Objective Optimization of Material Selection for Sustainable Products: Artificial Neural Networks and Genetic Algorithm Approach," *Materials & Design*, Vol. 30, No. 4, 2009, pp. 1209-1215.

[13] P. Zhou, B. W. Ang and D. Q. Zhou, "Weighting and Aggregation in Composite Indicator Construction: A Multiplicative Optimization Approach," *Social Indicator Research*, Vol. 96, No. 1, 2010, pp. 169-181.

[14] G. Seyfang, "Community Action for Sustainable Housing: Building a Low Carbon Future," *Energy Policy*, Vol. 38, No. 12, 2010, pp. 7624-7633.

[15] M. Malanca, "Green Building Rating Tools in Africa," In: *Conference on Promoting Green Building Rating in Africa*, Green Building Africa, Nairobi, 4-6 May 2010, pp. 16-25.

[16] L. Wastiels, I. Wouters and J. Lindekens, "Material Knowledge for Design: The Architect's Vocabulary, Emerging Trends in Design Research," *International Association of Societies of Design Research (IASDR) Conference*, Hong Kong, 16-19 July 2007.

[17] M. C. Quinones, "Decision Support System For Building Construction Product Selection Using Life-Cycle Management," A Thesis Presented to the Academic Faculty in Partial Fulfillment of the Requirements for the Degree Master of Science in Building Construction and Facility Management, Georgia Institute of Technology, Atlanta, 2011.

[18] W. B. Trusty, "Incorporating LCA in Green Building Rating Systems," Air & Waste Management Association, Ottawa, 2009.

[19] W. B. Trusty, "Sustainable Building: A Materials Perspective," Prepared for Canada Mortgage and Housing Corporation Continuing Education Series for Architects, 2003.

[20] W. B. Trusty, "Understanding the Green Building Toolkit: Picking the Right Tool for the Job," *Proceedings of the USGBC Greenbuild Conference & Expo*, Pittsburgh, 2003.

[21] W. B. Trusty, J. K. Meril and G. A. Norris, "ATHENA: A LCA Decision Support Tool for the Building Community," *Proceedings: Green Building Challenge '98—An International Conference on the Performance Assessment of Buildings*, Vancouver, 26-28 October 1998, p. 8.

[22] T. Woolley, "Natural Building: A Guide to Materials and Techniques," The Crowood Press Ltd, Ramsbury, Marlborough, Wiltshire, 2006.

[23] United States Green Building Council (USGBC), "LEED-Leadership in Energy and Environmental Design: Pilot Credit Library: Pilot Credit 1—Life Cycle Assessment of Building Assemblies and Materials," *US Green Building Council*, 2010.

[24] L. Florez, D. Castro and J. Irizarry, "Impact of Sustainability Perceptions on Optimal Material Selection in Construction Projects," *Proceedings of the Second International Conference on Sustainable Construction Materials and Technologies*, University Politecnica delle Marche, Ancona, Italy, Coventry University and The University of Wisconsin Milwaukee Centre for By-products Utilization, 28-30 June 2010, pp. 719-727. http://www.claisse.info/Proceedings.htm,

[25] L. Florez, D. Castro-Lacouture and J. Irizarry, "Impact of Sustainability Perceptions on the Purchasability of Materials in Construction Projects," *Proceedings of the* 2009 *ASCE Construction Research Congress*, Banff, 8-10 May 2010, pp. 226-235

[26] D. Castro-Lacouture, J. A. Sefair, L. Florez and A. L. Medaglia, "Optimization Model for the Selection of Materials Using the LEED Green Building Rating System," *Proceedings of the* 2009 *ASCE Construction Research Congress*, Seattle, Washington, 5-7 April 2009, pp. 608-617.

[27] E. Keysar and A. Pearce, "Decision Support Tools for Green Building: Facilitating Selection among New Adopters on Public Sector-Projects," *Journal of Green Building*, Vol. 2, No. 3, 2007, pp. 153-171.

[28] C. Bayer, M. Gamble, R. Gentry and S. Joshi, "AIA Guide to Building Life Cycle Assessment in Practice," The American Institute of Architects, Washington DC, 2010.

[29] ATHENA Institute, "The Impact Estimator for Buildings," 2011. http://athenasmi.org/tools/impactEstimator/

[30] ATHENA Institute, "The EcoCalculator for Buildings," 2011. http://athenasmi.org/tools/ecoCalculator/index.html

[31] Z. Kapelan, D. Savic and G. Walters, "Decision-Suppport Tools for Sustainable Urban Development," *Proceedings of the Institution of Civil Engineers, Engineering Sustainability*, Vol. 158, No. 3, 2005, pp. 135-142.

[32] S. Rahman, S. Perera, H. Odeyinka and Y. Bi, "A Knowledge-Based Decision Support System for Roofing Materials selection and Cost Estimating: A Conceptual Framework and Catamodelling," *25th Annual ARCOM Conference*, Nottingham, 7-9 September 2009, pp. 1-10.

[33] S. Rahman, S. Perera, H. Odeyinka and Y. Bi, "A Conceptual Knowledge-Based Cost Model for Optimising the Selection of Material and Technology for Building Design," In: A. R. J. Dainty, Ed, *24th Annual ARCOM Conference*, Association of Researchers in Construction Management, University of Glamorgan, 1-3 September 2008, pp. 217-225.

[34] E. Loh, T. Crosbie, N. Dawood and J. Dean, "A Framework and Decision Support System to Increase Building Life Cycle Energy Performance," *Journal of Information Technology in Construction*, Vol. 15, No. 2, 2010, pp. 337-353.

[35] G. K. C. Ding, "Sustainable Construction: The Role of Environmental Assessment Tools," *Journal of Environmental Management*, Vol. 86, No. 3, 2008, pp. 451-464.

[36] C. Hopfe, C. Struck, *et al.*, "Exploration of Using Building Performance Simulation Tools for Conceptual Building Design," *IBPSA-NVL Conference*, Delft, 20 October 2005, pp. 1-8.

[37] R. S. Perera and U. Fernando, "Cost Modelling for Roofing Material Selection," *Built Environment: Srilanka*, Vol. 3, No. 1, 2002, pp. 11-24.

[38] A. Mohamed and T. Celik, "An Integrated Knowledge-Based System for Alternative Design and Materials Selection and Cost Estimating," *Expert Systems with Applications*, Vol. 14, No. 3, 1998, pp. 329-339.

[39] M. A. A. Mahmoud, M. Aref and A. Al-Hammad, "An Expert System for Evaluation and Selection of Floor Finishing Materials," *Expert Systems with Applications*, Vol. 10, No. 2, 1996, pp. 281-303.

[40] K. Lam and N. Wong, "A study of the Use of Performance Based Simulation Tools for Building Design and Evaluation in Singapore," IBPSA, Kyoto, 1999.

[41] J. L. Chen, S. H. Sun and W. C. Hwang, "An Intelligent

Data Base System for Composite Material Selection in Structural Design," *Engineering Fracture Mechanics*, Vol. 50, No. 5-6, 1995, pp. 935-946.

[42] G. Soronis, "An Approach to the Selection of Roofing Materials for Durability," *Construction and Building Materials*, Vol. 6, No. 1, 1992, pp. 9-14.

[43] I. Giorgetti and A. Lovell, "Sustainable Building Practices for Low Cost Housing: Implications for Climate Change Mitigation and Adaptation in Developing Countries," Giorgetti and Lovell, South Africa, 2010.

[44] R. Ellis, "Who Pays for Green Buildings? The Economics of Sustainable Buildings," CB Richard Ellis and EMEA Research, New York, 2009.

[45] R. J. Cole, "Building Environmental Assessment Methods: Redefining Intentions and Roles," *Building Research and Information*, Vol. 35, No. 5, 2005, pp. 455-467.

[46] R. J. Cole, G. Lidnsey and J. A. Todd, "Assessing Life Cycles: Shifting from Green to Sustainable Design," *Proceedings*: International Conference Sustainable Building, Rotterdam, 22-25 October 2000, pp. 22-24.

[47] R. K. Yin, "Case Study Research: Design and Methods," 4th Edition, Sage Publications, Los Angeles, 2009.

[48] B. Reza, R. Sadiq and K. Hewage, "Sustainability Assessment of Flooring Systems in the City of Tehran: An AHP-Based Life Cycle Analysis," *Construction and Building Materials*, Vol. 25, No. 4, 2011, pp. 2053-2066.

[49] D. K. H. Chua, Y. C. Kog and P. K. Loh, "Critical Success Factors for Different Project Objectives," *Journal of Construction Engineering and Management*, Vol. 125, No. 3, 1999, pp. 142-150.

[50] T. L. Saaty, "Relative Measurement and Its Generalization in Decision Making Why Pairwise Comparisons Are Central in Mathematics for the Measurement of Intangible Factors the Analytic Hierarchy/Network Process," *RACSAM-Revista de la Real Academia de Ciencias Exactas, Fisicas y Naturales. Serie A. Matematicas*, Vol. 102, No. 2, 2008,

pp. 251-318.

[51] T. L. Saaty, "Time Dependent Decision-Making; Dynamic Priorities in the AHP/ANP: Generalizing From Points to Functions and from Real to Complex Variables," *Mathematical and Computer Modelling*, Vol. 46, No. 7-8, 2007, pp. 860-891

[52] T. L. Saaty, "Decision Making for Leaders: The Analytic Hierarchy Process for Decisions in a Complex World," RWS Publications, Pittsburgh, 2001.

[53] T. L. Saaty, "Fundamentals of the Analytic Hierarchy Process," RWS Publications, Pittsburgh, 2000.

[54] T. L. Saaty, "Fundamentals of Decision Making and Priority Theory with the Analytic Hierarchy Process," RWS Publishers, Pittsburgh, 1994.

[55] T. L. Saaty, "The Analytic Hierarchy Process," McGraw-Hill, New York, 1980.

[56] J. A. Alonso and M. T. Lamata, "Consistency in the Analytic Hierarchy Process: A New Approach," *International Journal of Uncertainty, Fuzziness and Knowledge-Based Systems*, Vol. 14, No. 4, 2006, pp. 445-459.

[57] P. Gluch and H. Baumann, "The Life Cycle Costing (LCC) Approach: A Conceptual Discussion of its Usefulness for Environmental Decision Making," *Building and Environment*, Vol. 39, No. 5, 2004, pp. 571-580.

[58] C. J. Kibert, "Sustainable Construction: Green Building Design and Delivery," 2nd Edition, John Wiley and Sons, Inc., Hoboken, 2008.

[59] R. Spiegel and D. Meadows, "Green Building Materials: A Guide to Product Selection and Specification," John Wiley & Sons, Inc., New York, 2010, pp. 1-7.

[60] M. F. Ashby and K. Johnson, "Materials and Design: The Art and Science of Material Selection in Product Design," Butterworth-Heinemann, Oxford, Boston, 2002.

[61] J. F. Hair, R. E. Anderson, R. L. Tatham and W. C. Black, "Multivariate Data Analysis," Prentice Hall, Upper Saddle River, 1998.

APPENDIX A: Feedbacks from Evaluators

The following are feedbacks and suggestions retrieved from users on the MSDS tool. The names of the participants were undisclosed to respect their anonymity.

"The system relates to issues concerned with local knowledge, local materials data, local climate know-how, local experts needed to operate system, which are hardly considered in other systems". I think it shows great promise and the mechanics are very well-developed and user-friendly,

"Material costs vary from location to location (especially in the USA where material costs vary not just from state to state but also from city to city". Perhaps when the material selection is sorted by the element choice,

this will seem more useful".

"It depends on what resources you are referring to; if referring to the underlying database, those are considerable. If referring to the resource needs of the organization that would use the model, not too costly to operate".

"The interface is very well-designed and easy to navigate. However, there is a need for more explanatory material to allow the user to understand what s/he is actually doing, and how to operate some parts of the model appropriately".

"In terms of its operation, interoperability, flexibility, usability and applicability, per se, it is very clear and straightforward; it's the underlying premise and data that needs little clarification in order for the user to operate the model effectively.

14

Architectural Features of Stilted Buildings of the Tujia People: A Case Study of Ancient Buildings in the Peng Family Village in Western Hubei Province, China

Kui Zhao[1*], William L. Tilson[2], Dan Zhu[3]

---

[1]School of Architecture & Urban Planning, Huazhong University of Science and Technology, Wuhan, China; [2]School of Architecture, University of Florida, Gainesville, USA; [3]Urban and Regional Planning, University of Florida, Gainesville, USA.

ABSTRACT

This paper describes and analyzes the stilted buildings of the Tujia people (an ethnic group living in mainland China), a distinctive building style unique to them, from the perspectives of site selection, spatial layout, construction techniques, and cultural inheritance. The cluster of stilted buildings (*Diaojiao Lou* in Mandarin Pinyin) in the Pengjia Village (meaning most of the villagers share the surname of Peng) is presented as a case study in this paper. The paper makes a case for their preservation as authentic carriers of the Tujia people's cultural history, which is quickly disappearing due to development pressures. Three preservation strategies are discussed to meet this preservation goal. The first is to provide a detail analysis of the construction language to guarantee authenticity in the documentation, preservation and restoration processes of the stilted buildings. The second is to keep alive the expert knowledge and skill of traditional artisans by involving them in the construction of new structures using *diaojiaolou* techniques. The third strategy is to encourage local people to "dress-up" discordant buildings constructed mid to late 20th century with well-mannered facades using traditional details such as suspension columns, *shuaqi*, and six-panel and bang doors. Taking as a whole, these strategies are presented to help local residents, preservation experts, developers and policy makers sustain the irreplaceable cultural heritage and economic independence of the Tujia people.

Keywords: Tujia People; Stilted Buildings; Ancient Architecture Surveying; Traditional Structural Features; Traditional Spatial Features

1. Introduction

Stilted buildings are unique to the Tujia people living in the mountainous region of western China, including Hubei Province, Chongqing municipality, Hunan Province, and Guizhou Province. They are typical architectural structures carefully adapted to the local ecology, environment, and geography, characterized by steep mountains and wood-covered topography, a moist and rainy climate, extremely hot summers, and severe winters [1,2]. The stilted buildings clearly represent the folk customs, and the artistic, cultural, and aesthetic preferences of the Tujia.

The stilted buildings in the Peng Family Village (Pengjia Village) in the mountains in Xuan'en County in the west of Hubei Province are the most typical representatives of such buildings [3]. The village is not easy to reach, and they are preserved in perfect condition due to their remote location. During the summer holiday of 2012, a team from Huazhong University of Science and Technology (HUST) surveyed the cluster of ancient stilted buildings hidden in the remote mountains in order to reveal the mystery of the Tujia Village.

2. Site Selection

The site selection of Pengjia Village represents the most intact cultural and building practices of all Tujia villages. There are more than 200 villagers in the 45 households in Pengjia Village [4]. Most of the villagers emigrated from

Architectural Features of Stilted Buildings of the Tujia People: A Case Study of Ancient Buildings in the Peng Family Village in Western Hubei Province, China

129

Hunan to Hubei Province by following the Youshui River, the most important river west of Hunan and Hubei Province. Most of the Tujia people live along the You-shui River, which they refer to as their "mother river" [5] At the end of the Qing Dynasty and during the 18th to the 20th centuries, the river was employed as the most important channel to transport salt from Sichuan Province to Hubei and Hunan Province [1]. Today, many elderly people still remember their experience of shipping salt to Pengjia Village. In the 200-year period when transporting salt was a major enterprise, there have been a few waves of immigration, which resulted from the growing population in the region. The immigrants maintained a primitive and self-sufficient way of life through farming and weaving; they lived in a closed region with little exchange and communication with the outside world aside from salt transport.

Along the banks of the Youshui River are more than a dozen Tujia villages such as the Wang Family Village, Zeng Family Village, Luo Family Village, Wu Family Village, and Baiguoba Village. The salt shipping and production are not only the pillar industry of the Tujia people, but also result in the popularity of the stilted buildings in this region.

Most importantly, the Peng Family Village has fostered the most beautiful and well preserved stilted buildings of the Tujia. The village lies on the south of the so-called Lotus Seat of the Goddess of Mercy (Kwan-Yin) at the foot of Kwan Yin Mountain. On the west of the village is a deep and long stream, over which there is a century-old wind-rain bridge (a local style of bridge that has a small structure built on the bridge to avoid wind and rain). The clean and transparent Longtan River (one of the tributaries of the Youshui River) flows through the village in its front section. On the Longtan River is a 40-meter-long and 0.8-meter-wide wood-board-paved cable bridge connecting the village to the outside world. Behind the village are steep hills and mountains covered by dense bamboo forests. Walking downstream along the Longtan River, you will witness the Lion Rock, Shuihong Temple and another village called Wangjia Village. The Pengjia Village and Wangjia Village both emigrated from Hunan province (**Figures 1-3**).

Viewed from afar, one is easily overwhelmed by the artistic glamour of the exquisite cluster of stilted buildings of the Peng Family village. Over nine buildings on piles stand on the front and rear sections of the village, which feature cornices, rake angles and traditional Chinese exterior decorations. There are also another dozen pillar-supported dwellings at the end of the stilted buildings closest to the mountain. The space in the pillars is

Figure 1. Distant view of the cluster of stilted buildings picture by Kui Zhao, 2013.

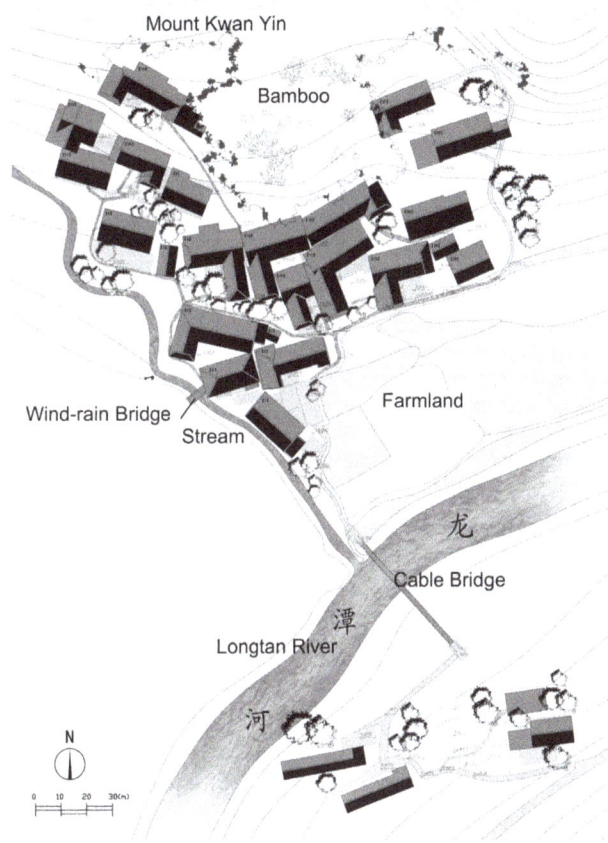

Figure 2. Site plan of Peng Family Village picture by Kui Zhao, 2012.

used as a passageway, warehouse, or stables and pens for cows and pigs. Most of the stairways and courtyards in the village are paved with precisely cut and well-maintained local slate. The stilted buildings and space in the courtyards are quite well-ventilated without the odors of the adjacent stables [6,7]. Even in summer, they provide a cool and dry environment, which is perfect for the moist and hot summer climate in western Hubei Province.

Figure 3. Stilted buildings of Peng Family Village drawing by Kui Zhao, 2012.

The Peng Family Village was built in front of the mountain, close to the water. The streams flowing on its sides form the borders of village. With the square shape, the village is the typical site selection of the Tujia people settlements.

3. Structural Features of Stilted Buildings

The stilted building is a kind of structure of through type timber frame that adapts to the topography in the mountain areas. Since there is an empty space in the lower level or slope of the hillside, the space is supported by many wooden columns that form the corridors under the huge roof and overhang balcony. The outmost columns are slender woods that are suspended from the roof and do not reach the ground. It seems that all the buildings are suspended by slender wood, which is the reason why

they are called stilted buildings. Though different from the ordinary pillar-supported buildings, the stilted buildings can still be labeled as special pillar-supported ones. We will explain the structural differences by taking as an example, the 3-dimensional anatomy model of a stilted building with the quasi-pavilion (**Figure 4**).

Figure 4. Construction process analysis computer modeling by Kui Zhao, 2007.

This stilted building is shaped like the letter L. It has the typical typology composed of one principal house and one wing. The foundation of the wing is lower than that of the principal house, and the lower level of the wing is suspended to form the quasi-pavilion. Some peripheral columns supporting the quasi-pavilion are not rooted on the ground. These columns are called step-supporting columns or suspension columns, whose weight is supported by the beams among the peripheral columns sitting on the floor or by the extrusions among the side columns [6,7]. The beams in the periphery of the quasi-pavilion are paved with wood boards to form the suspended corridor, at the end of which are the suspended short columns as the support of the corridor railing. These supports are called "Shuaqi". "Shuaqi" not only act as a support function, but also play an important role in decoration. The "Shuaqi" and the head of the suspension peripheral columns are shaped like balls or pumpkins, known as "head of Shuaqi" or "golden melon" by the local people. Because of their adjacency proximity to persons' viewport, the "golden melon" is one of the most important structural components of the decorations of Tujia buildings. The exterior sections of the square beam beyond the peripheral columns are called the "overhanging beams", which support the cornices. Because the cornice in the stilted buildings is often quite large, the supporting beams usually have two layers, forming the double-beams structure. The upper beam of smaller size is called the secondary beam, with the lower beam supporting the majority of the weight; thus it is called the primary overhanging beam. The primary beam often uses the naturally-bending trunk of large trees for the sake of weight holding. Sometimes the primary beam is shaped like a broadsword or a horse head. Thus, it is often called the "broadsword beam" or "horse head beam" [8]. The size and bending of the primary and secondary beams are significant for the gradient of the roof and design of the cornice (**Figure 5**).

Some Tujia buildings have transformed the double-beam structure into the "short-pillar structure" by adding a "short-pillar" on the overhanging beam, which the local people call a "stool pillar" [10]. On the ends of stool pillars are purlins that support the weight of the cornices. The primary overhanging beams go through the short-pillar and transmit part of the weight to the secondary small beams. Thus, the double beams and the short pillars collaborate to form a "stool pillar" to take more weight than the double beams do, making the force more rationally arranged. There are many other kinds of tectonic evolutions based on "double beams" and "stool pillar", such as "oblique beam" and "double pillar" [9, 10]. These designs have made the structure complex. Just like the "heads of Shuaqi", the ends of the "stool pillars"

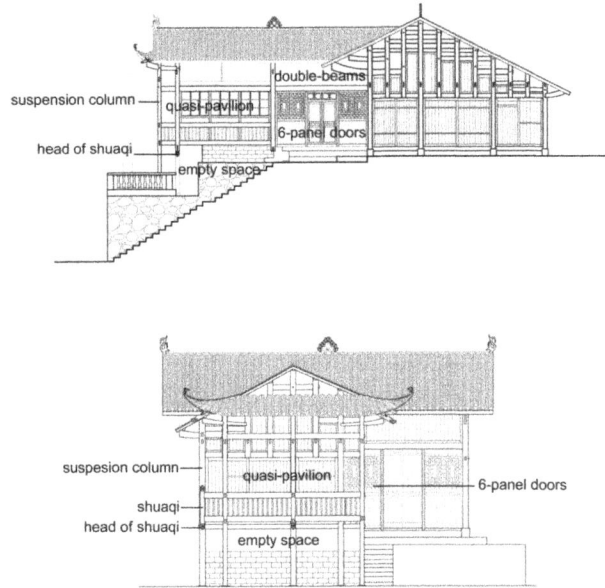

Figure 5. Façade map of the stilted building drawing by Kui Zhao, 2013.

are shaped in different designs and become the important decorations in Tujia buildings (**Figures 6** and **7**).

The quasi-pavilion, suspension peripheral columns, double-beams, stool-pillars, Shuaqi, handing columns, heads of Shuaqi and ends of hanging columns have become the most evident symbols of stilted buildings of the Tujia. The most distinctive scene of the stilted buildings in Pengjia Village is the row of quasi-pavilions along the foot of the mountain, presenting the most attractive and unique features of these buildings. Additionally, the cornice on the roof of the quasi-pavilions, catering to the elevation and light quality of the buildings, extrudes upward on the four corners and seems to be flying. These designs have made the façade highly animated and are typical of the Tujia buildings.

4. Details in the Buildings

4.1. Windows and Doors

The windows and doors in the stilted buildings in Pengjia Village are one of their most attractive features as serve as a tangible symbol of the Tujia people's wisdom and diligence in craft [10]. Though they are not as sophisticated and dignified as the windows and doors of the houses in Anhui Province, they are still known for their ancient, profound and diversified style, presenting the most delicate example of Tujia craftsmanship (**Figure 8**).

Most of the Tujia doors to the principal sitting room have six door panels that are 2.8 meters high and 5 meters wide. These six door panels, installed via the door spindles, form three doors to the room. The ends of each panel have the penetrating or relief flower-shaped

Figure 6. The decorations of stool pillars picture by Kui Zhao, 2011.

sculptures. In the middle section of the panel are the door windows of various designs. The 6-panel doors are sometimes fake. The genuine 6-panel doors can be opened forming three passages for people of different age and status in the family. During the Spring Festival when the villagers play the "lion lantern" [9,10], if the team fails to enter the doors following the proper etiquette, they will find it difficult to leave the room. The fake 6-panel doors, though they also have the same structure, have the panels on the sides simply fixed and not operable, leaving only two doors in the middle that can be opened. Some villagers would install two smaller door panels beside the 6-panel doors for the passage of chickens and dogs. The smaller door is 1.1 meter high and 1.7 meter wide. It is made of the timber of the *Cedrela chinensis* or "nut tree" [11,12]. Owing to the safe environment in the village, some houses are not equipped with the 6-panel doors and only have the smaller doors.

The secondary room is often equipped with only one wooden door with two panels. The other rooms use the single-panel door. There are two types of single-panel doors. One is the "embedded door" [9,10]. When closed, the door panel is perfectly imbedded into the door frame. The other is the "bang door" [9,10], because the door panel is larger than the door frame, and it will produce a "bang" noise when closing the door, which often results in the clash between the door panel and frame.

The windows are obviously used for lighting and ventilating; however, the windows in Pengjia Village have been given cultural content by the Tujia carpenters. These windows are shaped like Chinese characters. The door windows are shaped in a rectangle while the wall windows are square. The window designs are often symmetrical horizontally or vertically. The carpenters often make drawings first, then construct 3-cm patterns in a tenon-and-mortise design and connect the patterns to form the windows.

The window design reflects the craftsmanship and individuality of the Tujia carpenters and represents the pursuit of the Tujia people for a happy life. Every window design made of the patterns has its own meaning. Some carpenters even shape the patterns into sophisticated designs or animals. These designs are vivid and captivating even to those who do not understand their precise cultural meanings.

Unfortunately, there are only a few carpenters left in Pengjia Village who are trained in these traditional techniques. The owner of the house where our team lived was just such a carpenter. He lamented the loss of window carving techniques, saying most of the carpenters today have failed to inherit the traditional techniques and skill. Old carpenters make the windows with their own hands, but this distant village in the depth of mountains has been greatly influenced by the modern technologies. The young carpenters today mainly use machines to cut the battens, which are uniform in size and shape. However, when we measured the structural components of the ancient buildings, we found some components had different

Architectural Features of Stilted Buildings of the Tujia People: A Case Study of Ancient Buildings in the Peng
Family Village in Western Hubei Province, China

133

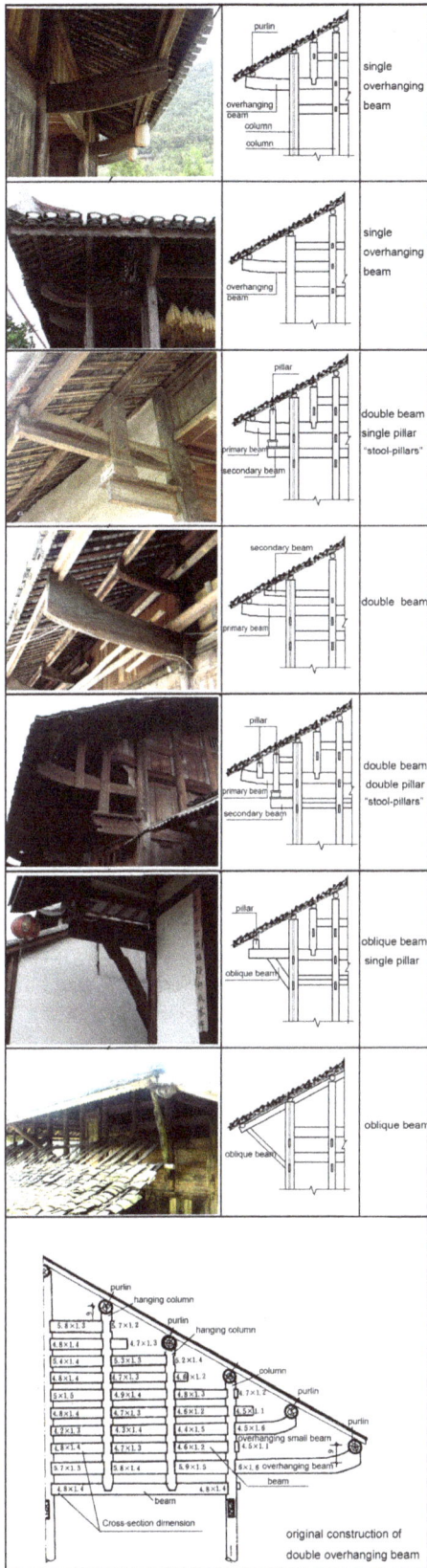

Figure 8. The real 6-panel door, the fake 6-panel door and the single-panel doors picture by Kui Zhao, 2007.

sizes. Probably the aesthetic attractiveness of the artisanship cannot be realized by the components made by the machines.

4.2. Roof

The building roofs in the Tujia villages produce an exquisite flowing visual effect. Seen from the vertical exterior layout, the buildings form the anatomy featured by touching the sky but staying away from the floor and the even top level but uneven floor level [11,12]. Such section planes are formed by adopting the techniques of suspended roof, omitted levels, and overlapping levels. As a result, viewers will sense the lively and vivid feeling without dullness or rigidity. The roof of the single stilted building is not complex in itself. It is often shaped like "—" or "L". Sometimes, the huge dark grey roof, the significant cantilever of the cornice, and the suspension space in the lower level will form the unstable composition of "heavy head and unstable feet". When the numerous facades are viewed in a cluster, however, the buildings become balanced, solemn, elastic, and rhythmic, producing a generous and profound aesthetic sense. If we see the overall layout of the Tujia stilted buildings, we will find them in an irregular and elastic cluster. Some houses are built catering to the topography of the mountain. Some produce overlapping layers of structure. Others are built on the edge of valleys. Many are lively and vivid, and a select number are sublime because of their positions on the hilltop.

Most of the stilted buildings in Pengjia Village are built at the foot of the mountain or hill. The narrow space under the cornices and the stairways following the ups and downs of the topography often produce the atmosphere of suddenly a village emerges in the eyes when people are wondering whether they have lost the directions [13,14]. Because of the large height difference in the site area, the large roof of the front building often surrounds the outdoor terrace of the rear building. Looking down from the suspended balcony of the higher building, you can see the overlapping and continuous roofs, looking like a rolling hill. These roofs seem to be surrounded by a crystal stream, a suspension bridge, yellow farm fields, and a huge, green mountain, which form fantastic rural scenery (**Figure 9**).

Figure 7. Tectonic evolution of double-beam structure; photo & drawing by Kui Zhao, 2008.

Figure 9. Roofs view of the cluster of stilted buildings picture by Kui Zhao, 2013.

4.3. Shrines

The Tujia buildings are cohabited by human beings and immortal beings [16,17]. The Tujia people must locate space in their homes to worship the immortal beings and their ancestors. These spaces are often set in the shrines or places equivalent to shrines, in the principal sitting rooms. Often sacred spaces are placed in the kitchen. People also believe that the immortal beings live in the stables, mills, workshops, or corners in the house [16,17]. In addition, different ethnic groups allocate different spaces in the house as shrines and adopt different functions and shapes for the shrines, which become an important symbol identifying the ethnic group (**Figure 10**).

The shrines in the Peng Family Village are often placed on the rear wall in the principal sitting room, in the middle of which is installed a wood board called a "shrine platform" to worship Grandfather Nuotuo and Grandmother Nuotuo believed to be the ancestors of Tujia people [15]. On the platform are placed the incense burner, candles, and straw paper. On the top of the shrine is another piece of wood board called a flame board, used to prevent against fire. Apart from the above-mentioned hardware in the shrine of the Tujia buildings, there is also the ancestral list describing the hometown and name of the ancestors pasted on the middle of the platform and the flame board. After everything is set, the priest of the Tujia people will be invited to hold ceremonies to usher in the immortals beings or ancestors into the shrine. After this ritual is completed, the space becomes a genuine shrine.

5. Spatial Features of the Buildings

The Tujia villages have the distinctive spatial forms composed of the narrow lane space in the village, the space under the cornices and the courtyard space surrounded by the roof, and the building and the environmental space beyond the village.

Figure 10. Shrine picture by Kui Zhao, 2007.

The villages of the Tujia people are often built on the river with a certain distance from the river; this distance can provide the buffer area when the flood comes. In addition, the farmland in the buffer area is fertile and becomes an excellent growing place for crops. The entrance roads to the villages are also built on the south bank of the river, making the river the natural protection for the villages. The sequence of the village layout is composed of the hill roads, river, suspension bridge, farmland, village, bamboo forests, and mountain in the background. Such a spatial layout has formed the diversified and complex exterior space of the village.

Walking into the village, crossing the winding lanes and stepping onto the stairways, you will enter the compact and diversified space in the single stilted buildings.

The stilted buildings have many forms; "—" shape, "L", and "U" shape are very popular. The Tujia people choose different styled dwellings according to the complex changes of the topographic landforms. They usually build dwellings parallel to the contour line of the mountain or hill, but still, many stilted buildings are built vertical to the contour line because of the limited site area in mountain rural area. The typical stilted building can be divided into two parts: one is set on a higher level, another on a lower area. The residential area in higher level has the sitting room and bedroom, and vertical to it is the suspended building. The suspension space in the lower level has the toilet, bathroom, and pigpen. The second floor, which connects to higher areas, has the dining room, kitchen room and another bedroom. By taking some typical stilted buildings in Pengjia Village as an example, we summarize their spatial features as follows (**Figures 11** and **12**).

Every building has many rooms, and every room is linked to each other. The sitting room is the most important space, which has many doors on the walls in each

Architectural Features of Stilted Buildings of the Tujia People: A Case Study of Ancient Buildings in the Peng
Family Village in Western Hubei Province, China

135

Figure 11. Analytical model of the single stilted building Modeling picture by Kui Zhao, 2010.

FORM	PLAN	SECTION OF TOPOGRAPHY
"–" shape parallel contour line		
"–" shape vertical contour line		
"U" shape parallel contour line		
"L" shape parallel contour line		
"L" shape vertical contour line		

Figure 12. Analysis of plane and section of stilted building drawing by Kui Zhao, 2010.

Figure 13. warm floor in sitting room picture by Kui Zhao, 2013.

direction. The other private rooms, such as the bedroom, often have two doors that provide access to it. It represents that the family is cohesive but does create an ambiguous awareness of privacy.

In winter, people flock together around a brazier (which is a large braze container in which charcoal is burned) in the center of the sitting room. Under the sitting room floor is empty, so the warm air flows into the empty space keeping the inside room warm. It is a simple but efficient folk technology (**Figure 13**).

Sewage is strictly separated from the hygienic areas. To make better use of the space in lower level, the stable and toilet that produce odor and sewage are often placed in this level. Thus, the wood structure of the higher level can remain dry and hygienic for the whole year. The stilted buildings also separate the inhabitants from the many insects and poisonous snakes living on the hill

slopes. The courtyard serves as the transit space for transport of goods. The rooms are arranged on the hill slope, and consequently, they may not be reached by directly by walking. The courtyard is often used as a temporary storage space. The 2 meters wide stone paving is beyond the extended cornice, and the cornices are used as the shelter against rain and strong sunlight when people walk on them.

6. Conclusions

The Tujia people's stilted buildings have their own ethnic distinctiveness in construction, such as the quasi-pavilion, suspension peripheral columns, double-beams, stool-pillars, and huge roof, balcony and cornices. The most distinctive feature is that many wooden pillars, which help the inhabitants adapt to living in mountain environment, support the buildings. High above the ground, stilted buildings have the following advantages:

First, it can keep people away from deadly dangers, such as miasma, poisonous vegetation, venomous snakes, and huge wild animals.

Second, people can stay away from the humidity close to the ground and prevent humidity related diseases.

Finally, there is better lighting upstairs, so people can work on delicate handcrafts or simply enjoy the light.

The Tujia people also create their own architectural decorative art: "Shuaqi", handing columns, heads of Shuaqi and ends of hanging columns, 6-panel doors, and carved patterns windows. All of these have become striking characteristics of the stilted buildings of the Tujia people.

The carefully preserved stilted buildings in Pengjia Village have inherited the traditional features of the Tujia people's architectures. Based on a large number of our first-hand information through field research, ancient architecture surveying, and mapping in this village, in combination with our research in the Tujia areas over the past decade, the paper aims to record the real history and keep the local art and traditional technology.

Since the end of last century, rapid economic development in the past 20 years in China has resulted in the introduction of cheap undifferentiated concrete buildings in the Tujia area. Residents are faced with financial and natural resources challenges such as decreasing forests continuous rise of timber prices, and the rapidly dwindling number of skilled wood workers—conditions that force them to abandon traditional buildings. Additionally, the existing wooden structures need regular maintenance, such as having tiles replaced and being brushed with tung oil. Since large populations are migrant workers, many of the houses on stilts become empty nests. Without proper care, the stilted houses naturally collapse very easily. This constantly required care is what makes people give up on the stilted buildings. Local residents now tend to build simple concrete buildings with low costs rather than stilted buildings with complex wooden structures, thus the regional characteristics of the Tujia architecture are gradually disappearing.

In the past five years, highway extensions and railway construction have brought large number of tourists to the Tujia area [15,17]. Visitors revel in the beautiful natural scenery, while simultaneously marveling at these stilted building clusters integrated with the landscape. This has led the government to focusing on the return of traditional building methods in an attempt to attract more visitors in order to meet the tourism demands and promote economic development in the Tujia villages such as Pengjia.

For example, starting in 2008 in the EnshiTujia Autonomous Prefecture, Hubei Province, the government began to restore the stilted buildings gradually from three aspects to maintain the rural traditional regional characteristics.

The first priority is to protect the integrity of ancient villages, such as the Pengjia Village, as articulated in this paper. Our team for example has performed measured drawings, photographed every ancient building in the village, established original files for them, set up protection signs, and stationed protection mechanisms to protect the village. Demolition, reconstruction, and new building construction are strictly prohibited in the ancient village, and special funds will be allocated for the repairing and reinforcement of these irreplaceable cultural resources. To protect the ancient architecture, special attention has also been paid to the village environment and village culture, e.g., the restoration of riverine and mountain vegetation, and support of the traditional dances and customs of Pengjia Village as intangible cultural heritage. Such examples include the "Hands Waving Dance", "Drum Melody for Weeding", and "Xuanen Play". Tourists are invited to participate in the dances to experience the true traditional culture and meaning of the dances (**Figure 14**).

The second priority is to construct new buildings in the traditional way. The construction of new villages and expansion of existing villages in Tujia area require planning and construction following traditional ideas, which is completely out of the ordinary compared to commercial development modes. Planners need to extract and recombine traditional elements based on meticulous research on traditional Tujia villages (such as the windows, doors, roof, balcony, shrines and other elements as mentioned in the paper) and cooperate with the traditional woodworkers. For example, our research team has made numerous explorations and conducted various experiments in the design of the Pengjia Village Visitor Center, Qingyang Dam Ancient Village Renovation, Yumuzhai Ancient Village Planning (**Figure 15**). In addition, we highly encouraged the local residents to participate in together with the Tujia building construction professional team. Using these approaches, we hoped to encourage the residents to consciously build and maintain the traditional wood structure buildings. In the meantime, the local government pays subsidies on the increased cost causing by building the complexity of the stilted buildings.

The third priority is to restore and apply the traditional style onto discordant architecture. The local government describes it as "dressing up" the building, and this is mainly targeted at the large number of newly built rough concrete buildings at the end of last century. People have begun to add wooden roof structures at the top of the concrete structure, fitted them with wooden battens for the exterior wall, replaced the concrete balcony railings with suspension columns, Shuaqi, and head of Shuaqi used in the balcony of Tujia people, and replaced the aluminum alloy doors and windows with unique Tujia six-panel doors and bang doors. With this treatment, the exterior of the buildings that cannot be removed now has a traditional cover and is harmonious with the surrounding ancient villages (**Figure 16**).

We are applying our research through active involvement in the protection of settlements with Tujia characteristics and construction practices. The purpose of this ongoing initiative is to preserve the regional characteristics

Figure 14. Protection of traditional buildings and folk customs picture by Kui Zhao, 2011.

Architectural Features of Stilted Buildings of the Tujia People: A Case Study of Ancient Buildings in the Peng
Family Village in Western Hubei Province, China

137

Figure 15. Directing Tujia people to build new stilted buildings using authentic traditional techniques picture by Kui Zhao, 2010.

Figure 16. "Dress up" the discordant architecture picture by Kui Zhao, 2011.

of Tujia buildings and to help the international community understand the unique architectural forms of this ethnic group in inland China.

Ultimately, it is the inherent beauty of traditional Tujia architecture that demands that we share this research to protect and preserve the Tujia villages for future generations.

7. Acknowledgements

The research was sponsored by the National Natural Science Foundation of China (No. 50978111).

REFERENCES

[1] K. Zhao, "Sichuan Salt Trail—Architecture and Village in the Field of Vision of the Cultural Route," Southeast University Press, Nanjing, 2008.

[2] D. Luo and K. Zhao, "Southwest Residential," Tsinghua University Press, Beijing, 2009.

[3] K. Zhao and M. Wan, "Hubei Xuan'enPengjia Village," *Journal of Urban Planning*, Vol. 3, No. 8, 2009, pp. 50-52.

[4] M. F. Zhong, "Research on Traditional Stone Buildings in West Hunan Rural Areas," *Advanced Materials Research*, Vol. 450, No. 2, 2012, pp. 218-222.

[5] C. Zhou, "Spatial Features of the Traditional Vernacular Dwelling in Three Gorges Area," *Journal of Anhui Agricultural Sciences*, Vol. 25, No. 3, 2011, pp. 122-124.

[6] X. H. Liu and J. Jia, "Sustainable Development of Tujia People's 'Diaojiaolou' House in Southwest of Chongqing," *International Conference on Multimedia Technology (ICMT)*, Hangzhou, 26-28 July 2011, pp. 1261-1264.

[7] J. Zheng and J. Yu, "Preservation and Utilization of Traditional Inhabited Areas in the North of Guangxi—With Ping'an Village in Longsheng County, Guilin City as an Example," *Planner*, Vol. 8, No. 1, 2006, pp. 56-58.

[8] Y. H. Lin, "New Rural Construction Tujia Stilted Development and Application," Nanjing Arts Institute (Art and Design), Nanjing, 2010, pp. 6-12.

[9] Q. Lu, "Tujia Residential Characteristics and Formation," *Huazhong Architecture*, Vol. 2, No. 3, 1996, pp.69-74.

[10] F. Peng, "Buildings on Stilts to Peng ZhaiTujia Traditional Houses about the Value and Protection," *Journal of Development of Small Cities & Towns*, Vol. 1, No. 2, 2011, pp. 2-7.

[11] Y. Sun and T. Lin, "Building Technology and Art of Tujia Houses in the Southeast Chongqing," Chongqing Jianzhu University, Chongqing, 2006, pp. 2-5.

[12] Y. Wang and Y. Zheng, "Western Hubei TujiaXuan'en Characteristics and Protection of Residential Houses on Stilts," *Huazhong Architecture*, Vol. 7, No. 5, 2005, pp. 31-35.

[13] M. Y. Xiang, Y. K. Yuan and Y. Huang, "Tujia Architectural Style Remediation Practices: Case Study of Qianjiang Ancient Water Town," *Art and Design*, Vol. 19, No. 7, 2009, pp. 79-83.

[14] Y. H. Xin and B. Luo, "On the Cultural-Permeation between Tu and Han Nationality from the Perspective of Wood-Carving in Furniture of Wulin Religion," Huazhong Normal University (Humanities and Social Sciences), Wuhan, 2004, pp. 10-22.

[15] Y. Q. Yao, "Aesthetical Study of Tujia Residential Architectures," *Shanxi Architecture*, Vol. 1, No. 7, 2004, pp. 13-17.

[16] W. Zhao, "Three Gorges Residential Courtyards Study—An Analysis Reservoir Republican Courtyard Space Morphology Evolution," Chongqing Architecture University, Chongqing, 2000, pp. 100-108.

[17] Y. P. Zhou and Q. Y. Li, "Analysis of Cultural Integration in the Evolution of Western Hubei Tujia Houses," *Chinese and Overseas Architecture*, Vol. 3, No. 3, 2012, pp. 23-31.

15

Quality Study of Automated Machine Made Environmentally Friendly Brick (KAB) Sample Using Film Neutron Radiography Technique

Khurshed Alam[1*], Robiul Islam[2], Sudipta Saha[1], Nurul Islam[1], Syed Azharul Islam[2]

[1]Institute of Nuclear Science and Technology, AERE, Savar, Bangladesh; [2]Department of Physics, Jahangirnagar University, Savar, Bangladesh.

ABSTRACT

Neutron radiography (NR) technique has been adopted to study the internal structure and quality of the KAB bricks made by Hoffman kiln method. Thermal neutron radiography facility installed at the tangential beam port of 3 MW TRIGA Mark-II Research Reactor, AERE, Savar, Dhaka, Bangladesh is used in the present study. Measurements were made to determine the internal structure and quality of the automated machine made environmentally friendly brick sample. In this case, optical density/gray values of the neutron radiographic images of the sample have been measured. From these measurements, the porosity, water penetrating height, water penetrating behavior, initial rapid absorption of water (IRA), elemental distribution/homogeneity and incremental water intrusion area in the sample have been found. From the observation of different properties, it is seen that, homogeneity of the Hoffman kiln brick KAB is not perfectly homogeneous and contains small internal porosity; the incremental water intrusion area is very poor, and the water penetrating height through the two edges is higher than the middle part; the initial rapid absorption (IRA) rate is also very poor and the water penetrating behavior of the samples is different as like as stair, capillary, wave and zigzag shape. From these points of view, it is concluded that the quality of the environmentally friendly brick KAB is better. The results obtained and conclusion made in this study can only be compared to the properties of bricks produced under similar conditions with similar raw materials.

Keywords: Neutron Radiography Technique; Water Penetrating Height/Behavior; IRA

1. Introduction

Neutron Radiography (NR) is a technique of making a picture of the internal details of an object by the selective absorption of a neutron beam by the object. NR uses the basic principles of radiography whereby a beam of radiation is modified by an object in its path and the emergent beam is recorded on a photo film (detector). In general, the radiography technique is nothing but a simple process of exposing some objects to an X-ray, gamma-ray, neutron beam and some other types of radiation and then attenuated outgoing beam from the object is passing through a special type of photographic film to form images of the objects on the radiographic film or detector. Also it is called a non-destructive testing (NDT) [1] and

evaluation technique of testing non-nuclear and nuclear materials and industrial products. NR is an imaging technique which provides images similar to X-ray radiography and complementary technology for radiation diagnoses. Neutron radiograph gives the information of the internal structure of an object; it can detect light elements, which have large neutron absorption cross-sections like hydrogen and boron; it is completely complementary to other NDT techniques, like X-ray or gamma-ray radiography. The atoms of the object material scattered or absorbed the radiation and so the beam reaching the detector shows an intensity/gray value pattern representative of the internal structure of the object [2]. Any in-homogeneity in the object on an internal defect (such as voids,

Quality Study of Automated Machine Made Environmentally Friendly Brick (KAB) Sample Using
Film Neutron Radiography Technique

139

cracks, porosity, inclusion, corrosion etc.) and morphological change in the plant pod seeds [3] will show up as change in gray value/radiation intensity reaching the detector. Under these techniques, detecting faults in neutron shielding materials, flow visualization: real time neutron radiography, quality control of explosive devices, defects in ceramics materials, aircraft component, surface corrosion on aluminum, medical and biological applications, investigations of the root soil system, migration/rising of water in various building products/building materials, physical description of water transport in a porous matrix of the sample material, density fluctuations and porosity detection in ceramics etc [4-20]. Clay is a widely available raw material that survives very well in its fired form. Clay brick has been found in the ruins of ancient civilizations [21]. Bhatnagar *et al.* saw that properties of these bricks are affected as a result of physical, chemical and mineralogical changes [22,23]. Mbumbia *et al.* investigated that compressive strength and water absorption are two major physical properties of brick that are good predictors of bricks ability to resist cracking of face [24]. Few scientists studied that compressive strength is highly affected by firing temperature method of production, and physical, chemical and mineralogical properties of the raw material [22,25]. Water absorption is a measure of available pore space and is expressed as a percentage of the dry brick weight. It is affected by properties of clay, method of manufacturing and degree of firing. Some of the researchers studied that firing shrinkage increases with higher temperatures [26]. The quality depends on the firing temperature and firing time also. Decreasing firing temperature and shortening firing time do not only reduce the cost of production but also increase the productivity of the factory.

Environmental concerns have been raised in some parts of the world where coal is the main power generating sources and where bricks are also the main building material. Most of the scientists believe that fly ash on its own can be an excellent raw material for brick making. This has now been proven and a patent is taken for the manufacture of bricks from fly ash [27].

Many ancient cultures have made useful decorative items such as pottery, figurines, building tiles, and burial containers that become important parts of the archaeological record. The material aspects of clay and ceramic technology, the physical properties of clay and various firing methods can be investigated using archaeometric techniques [28,29]. Properties of bricks are affected as a result of physical, chemical and metrological changes [23,30]. Water absorption is a measure of available pore space and is expressed as a percentage of the dry brick weight. It is affected by properties of clay, method of

manufacturing and degree of firing. Water absorption capacity of the brick affects the surface finishing of the brick-laid wall [21,26,31]. Ancient technologists and archaeological material researchers have employed standard techniques such as X-ray radiography, X-ray diffraction (XRD), scanning electron microscopy (SEM), and neutron activation analysis (NAA) to study structure and composition of ceramic materials [28,29]. Neutron radiography has been used to detect internal defects in some materials such as ceramics [9], tiles [10] and different building industries [11]. The technique is also adopted for the study of water absorption behavior in biopol, jute-reinforced-biopol composite [12] and wood plastic composites [13] etc. In the present work, neutron radiography technique has been adopted to the determination of elemental distribution/homogeneity, porosity, incremental intrusion area of water/water penetrating height and penetrating shape/behavior, and initial rapid absorption (IRA) of water in the sample as well as the quality of automated machine made environmentally friendly KAB brick.

2. Experimental Facility

The experimental neutron radiography facility installed at the tangential beam port of 3 MW TRIGA Mark II reactor in the Institute of Nuclear Science and Technology, Atomic Energy Research Establishment, Savar, Dhaka, Bangladesh. The neutron radiography facility consists of the following devices/equipment.

2.1. Bismuth Filter

In the NR facility at TRIGA reactor of BAEC a 15 cm long Bi filter in the tangential beam port is used to reduce the intensity of gamma ray significantly from the beam to prevent the unwanted fogginess in the radiographic image.

2.2. Cylindrical Divergent Collimator

A cylindrical divergent collimator made of 120 cm long aluminum hollow cylinder with 5 cm and 10 cm diameter at the inner and outer end, respectively, has been inserted in the tangential beam port to collimated neutron beam of the reactor. The advantage of the divergent collimator is that a uniform beam can be projected easily over a large inspection area. Collimators are required to produce a uniform beam and thereby produce adequate image resolution capability in a neutron radiography facility.

2.3. Lead Shutter

The outer end of the tangential beam tube is equipped with a lead-filled safety shutter and door to provide limited gamma shielding. The thickness of lead in the shutter is 24 cm and the diameter of the shutter is 33 cm.

2.4. Beam Stopper

A wooden box with dimension of 68 cm × 40 cm × 68 cm has been made with the attachment of four ball bearings on the bottom part of it for forward and backward movement in front of the tangential beam port. It looks a wooden box, which contains neutron-shielding materials like paraffin wax and boric acid in 3:1 ratio by weight for neutron shielding.

2.5. Sample and Camera Holder Table

There is a sample and camera holder table with both horizontal and vertical movement facility placed in front of the beam line.

2.6. Beam Catcher

To absorb transmitted and scattered neutron and gamma radiations a beam catcher with dimension 100 cm × 100 cm × 85 cm has been placed behind the sample and camera holding table. A 30 cm × 30 cm × 30 cm hole has been made in the middle of the front face of the beam catcher which coincides with the central axis of the beam port. A 30 cm × 30 cm × 15 cm lead block weighing 125 Kg has been placed at the back side of the hole for

gamma shielding. For neutron shielding a mixture of paraffin wax and boric acid has been used in the catcher. The total weight of the beam catcher is 968 Kg.

2.7. Biological Shielding House

The emitted neutron and the gamma rays are extremely dangerous for human body. This is why, to prevent these harmful rays to spread over the entire environment a biological shielding house has been built around the NR facility of the tangential beam port. It is made of special concrete containing cement, heavy sand (magnetite, ilmenite and ordinary sand) and stone chips in the ratio 1:3:3. Paraffin wax and boric acid in 3:1 ratio by weight were also used inside the biological shielding wall for neutron shielding. The width and height of the biological shielding wall of the facility are ≈ 3.0 ft and 6.5 ft, respectively. Details of the NR facility can be found elsewhere [3,32,33]. The schematic diagram of the neutron radiography facility of 3 MW TRIGA Mark II Reactor, AERE, Savar, Dhaka is shown in **Figure 1**.

3. Experimental Procedure

3.1. Collection, Preparation and Size of the Sample

Sample has been collected from Kapita auto bricks limited located at Joypura, Dhamrai, Dhaka, Bangladesh.

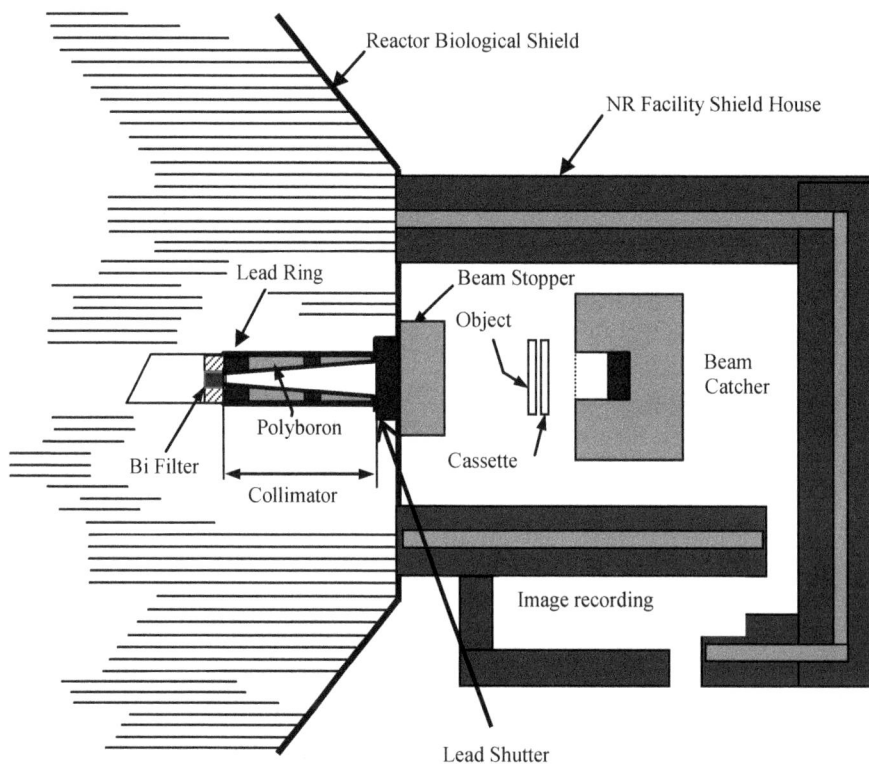

Figure 1. Schematic diagram of the neutron radiography facility.

For final preparation, the sample is polished manually by using series paper, cement block, diamond cutter, and then the sample was dried at daylight/dryer machine until to get the constant weight. The sample is the rectangular shape and its size is $23.000 \times 11.360 \times 6.540$ cm^3 and $23.050 \times 10.821 \times 6.480$ cm^3 for KAB 1 and KAB 2, respectively. In the case of KAB 2 sample, coal is mixtures with the soil and this coal is used to burn it. But in case of KAB 1 sample, coal is used into the brick kiln to burn the sample.

3.2. Loading Converter Foil and Film in the NR Cassette

A thin converter (gadolinium metal foil of 25 μm thickness) was placed at the back of the X-ray industrial film. The loading of the X-ray industrial film (Agfa structurix D$_4$DW) into the NR cassette (18 cm × 24 cm) is a simple procedure [14]. There are a number of steps to place the industrial X-ray film into the NR cassette to protect the film against daylight and lamplight.

3.3. Placing of Sample and the NR Cassette

The sample is placed in close contact with the NR cassette and directly on the sample holder table. The NR cassette is placed on the cassette holder table. Both of NR cassette and sample are placed in front of the neutron beam having 30 cm in beam diameter.

3.4. Determination of Neutron Beam Exposure Time

Exposure means passing of neutron beam through a sample and holding it onto a special film (X-ray industrial film) in order to create a latent image of an object in the emulsion layers of that film. Exposure time differs for different samples, depending on the intensity of the neutron beam, density and thickness of the sample and neutron cross-section. The optimum exposure time of the sample was determined by taking a series of experiments/radiographs at different exposure time, while the reactor was operated at 250 KW. For the present experiment we found the optimum exposure time is 60 minutes. The sample was then irradiated for that optimum time to obtain good neutron radiographs.

3.5. Immersion Procedure of the Brick Sample

The brick sample is placed in a plastic pan and a constant 2.0 cm height of water level is maintained. The water level is observed very carefully and adds extra water to maintain water level at 2 cm during the immersion time. After time of interest (TOI) such as 5, 10, 15 and 20 minutes brick sample take off from the pan and extra water

of out side the sample is removed by using the tissue paper.

3.6. Obtained Radiographic Images of the Sample

3.6.1. Irradiation
While all the procedures (a-e) were performed, the neutron beam was disclosed by removing the wooden plug, lead plug and beam stopper from the front side of the collimator. Each sample was then irradiating for the optimum time (60 min) one by one at various immersion time.

3.6.2. Developing
Developing is an image processing technique by which the latent image recorded during the exposure of the material is converted into a silver image [34]. Developing process is completed at 20°C for 5 minutes.

3.6.3. Fixing
When the developing is completed a conventional photographic material must be treated in an acid stop bath or it must be rinsed in water, after which it is treated in a fixation bath. The fixation solution will dissolve the unexposed silver-halide crystals leaving only the silver grains in the gelatin. The fixing is completed with in 5 minutes and controls the fixture temperature at 20°C.

3.6.4. Washing
In between developing and fixing the radiographic film, it is necessary to wash for 1 minute at flowing tap water.

3.6.5. Final Washing
The silver compound which was formed during the fixing stage must be removed, since they can affect the silver image at the latter stage. For this reason the film must be washed thoroughly in flowing tap water for 15 minutes after completion of developing and fixing process.

3.6.6. Drying
After the final washing, the films were dried by clipping in a hanger at fresh air/or in a drying cabinet.

After developing, washing, fixing and the final washing obtained radiographic images (**Figures 2** and **3**) of the required KAB brick sample at different immersion time.

4. Mathematical Formulation

4.1. Optical Density Measurement

The neutron intensity before reaching the brick sample (object) is different from the intensity of the neutron after passing through it. The relationship between these two intensities is expressed through the following equation [15]

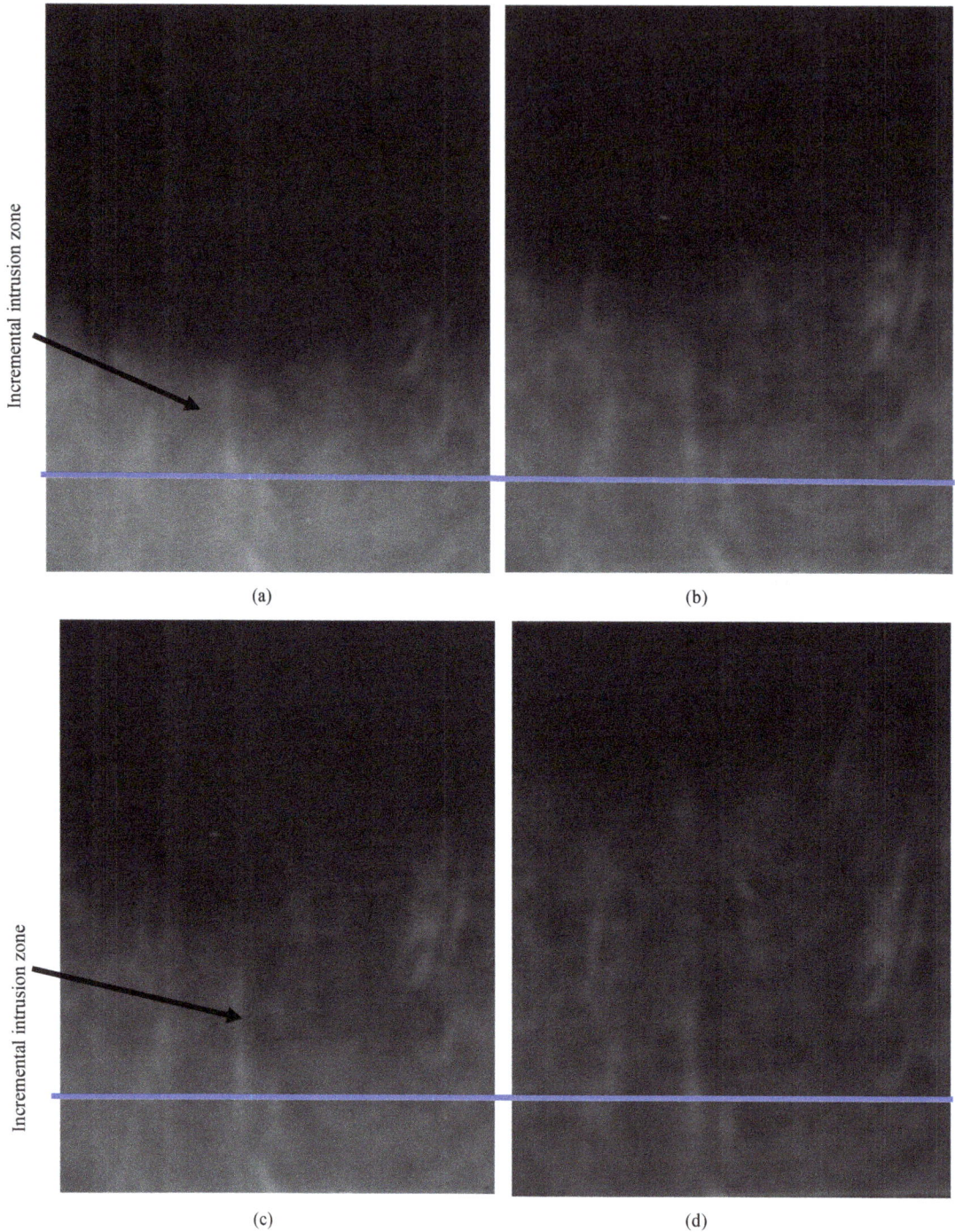

Figure 2. NR images of KAB 1 for (a) 5 min and (b) 10 min water absorption. NR images of KAB 1 for (c) 15 min and (d) 20 min water absorption.

$$I = I_0 e^{-\mu x} \qquad (1)$$

where, e = base of natural logarithms, x = thickness of an object, μ = linear neutron attenuation coefficient, I and I_0 are the neutron intensity after passing through the object and the neutron intensity incident on the object, respectively.

The mathematical expression for the optical density [16] at a point of the film/NR image, D is given by:

$$D = \ln\left(A_0 / A\right) \qquad (2)$$

Here, A_0 = response of densitometer without the sample image and A = response of densitometer with the sample image.

Quality Study of Automated Machine Made Environmentally Friendly Brick (KAB) Sample Using
Film Neutron Radiography Technique

143

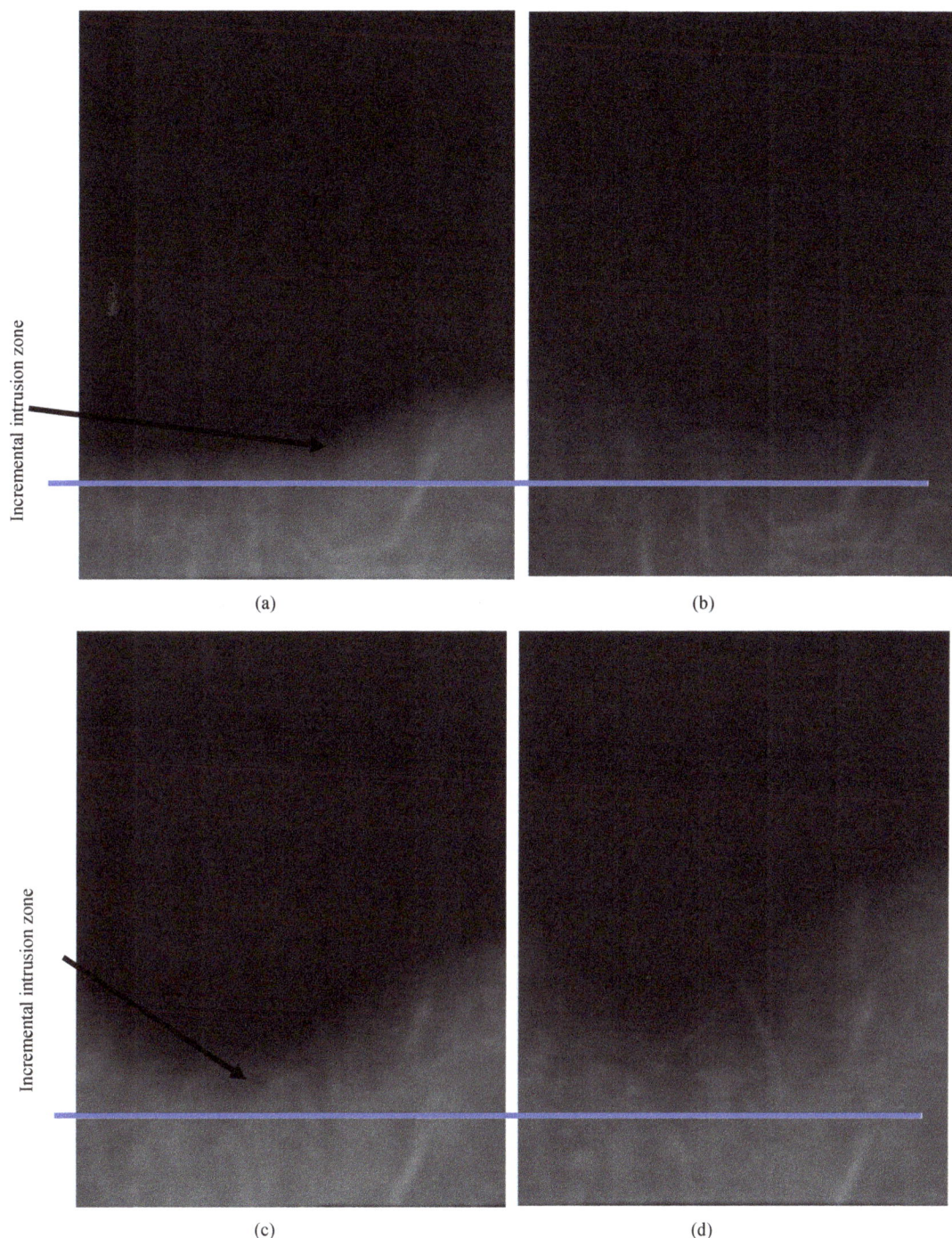

Figure 3. NR images of KAB 2 for (a) 5 min and (b) 10 min water absorption. NR images of KAB 2 for (c) 15 min and (d) 20 min water absorption.

The density of film is measured with an optical densitometer (Model 07-424, S-23285, Victoreen Inc. USA) [5]. A small beam of light from the light source passes through the film area which is measured by densitometer. On the other side of the film, a light sensor (photocell) converts the penetrated light into an electrical signal. A special circuit performs a logarithmic conversion on the signal and displays the results in density units. The primary use of densitometers in a clinical facility is to monitor the performance of film processors. Actually, optical density is the darkness, or opaqueness, of a transparency film and is produced by film exposure and chemical processing. An image contains areas with different densities that are viewed as various shades of gray.

4.2. Gray Value

The visual appearance of an image is generally characterized by two properties such as brightness and contrast. Brightness refers to the overall intensity level and is therefore influenced by the individual gray-level (intensity) values of all the pixels within an image. Since a bright image (or sub image) has more pixel gray-level values closer to the higher end of the intensity scale, it is likely to have a higher average intensity value. Contrast in an image is indicated by the ability of the observer to distinguish separate neighboring parts within an image. This ability to see small details around an individual pixel and larger variations within a neighborhood is provided by the spatial intensity variations of adjacent pixels, between two neighboring sub images, or within the entire image. Thus, an image may be bright (due to, for example, overexposure or too much illumination) with poor contrast if the individual target objects in the image have optical characteristics similar to the background. At the other end of the scale, a dark image may have high contrast if the background is significantly different from the individual objects within the image, or if separate areas within the image have very different reflectance properties.

Although the intensity distribution within any real-life image is unlikely to be purely sinusoidal, these definitions provide a basis for comparison. For example, an image that contains pixels with brightness values spread over the entire intensity scale is likely to have better contrast than the image with pixel gray-level values located within a narrow range. The relationship between the intensity spread at the pixel level and the overall appearance of an image provides the basis for image enhancement by gray-level transformation. The terms gray value and intensity are used synonymously to describe pixel brightness. Actually, the specific relationship between the shades of gray or density and exposure depends on the characteristics of the film emulsion and the processing conditions. This gray value is measured using image analysis software Image J [35].

5. Results and Discussions

In the present investigation NR techniques has been adopted to study internal defects such as in-homogeneity, porosity/voids, initial rapid absorption (IRA), water penetrating rate/behavior and incremental intrusion area of automated machine made environmentally friendly KAB bricks. Automated machine made environmentally friendly bricks industry (made by Hybrid Hoffman Kiln method) is established very recently in Bangladesh. The NR techniques allowed us to comment on the quality of this type of brick samples from the measurement of the gray value/optical densities of their neutron radiographic images.

5.1. Porosity/Voids and Homogeneity of the Samples

The quality of a brick samples depends on the proper distribution of the contents, porosity, hardiness, water absorption behavior etc. in the sample. In this section, porosity, elemental distribution of the samples has been studied by measuring gray value/intensity from the neutron radiographic images of each sample. Variation of gray values of the radiographic images of the samples indicates that the constituent components of the samples are not uniformly distributed and having internal porosity.

The **Figure 4** shows the gray value versus pixel distance plots of radiographic image of the KAB sample. The gray value has been obtained by drawing line profile of 1056×1600 pixel area on the radiographic images of an object. From this figure it is observed that in most of the places the variation of gray value is not regular manner for KAB 1 but in few places it is regular. It is also observed that KAB 1 sample is not perfectly homogeneous and contains little porosity because of irregularity of gray value. In the same figure for KAB 2, it is observed that variation of gray value in most of the pixel point is slightly irregular in nature. This shows that most of the regions for KAB 2 is homogeneous and small region is inhomogeneous. Small variation of gray value/intensity indicates the presence of less internal porosity of that place/area.

5.2. Water Penetrating Height at Different Immersion Time of the Samples

KAB 2

Water penetrating/rising behavior of the KAB 2 sample at different immersion time such as 5, 10, 15, 20 minutes is shown in **Figure 5**. From these graph it is observed that due to 5 minutes immersion water rises in upward direction is 2.6 cm and 4 cm through two edges and 3 cm at the middle side. In case of 10 minutes, penetrating of water at the middle place is 4 cm and through the edges this penetration is about 6 - 6.5 cm. For 15 minutes water immersion, the water uptake is 4 cm at middle and at two edges the water uptake is 6.4 - 6.6 cm. For 20 minutes, water uptake is 5 cm at middle and at two edges is 7 - 8 cm. From above investigation it shows that at first 5 minutes the water uptake through the middle is very higher than that of 10, 15 and 20 minutes. Except for first 5 minutes water immersion, water rises through the two edges is higher than the middle part.

KAB 1

Quality Study of Automated Machine Made Environmentally Friendly Brick (KAB) Sample Using
Film Neutron Radiography Technique

145

Homogeneity Measurement

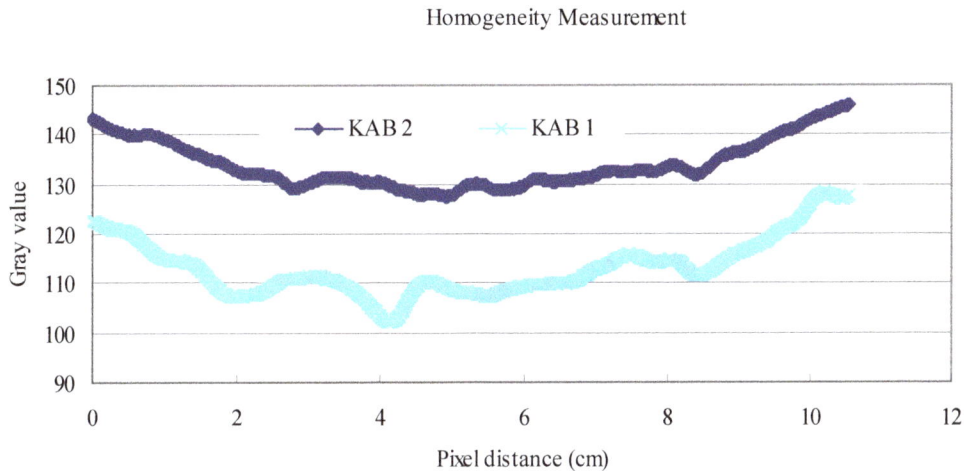

Figure 4. Gray value vs. pixel distance curve.

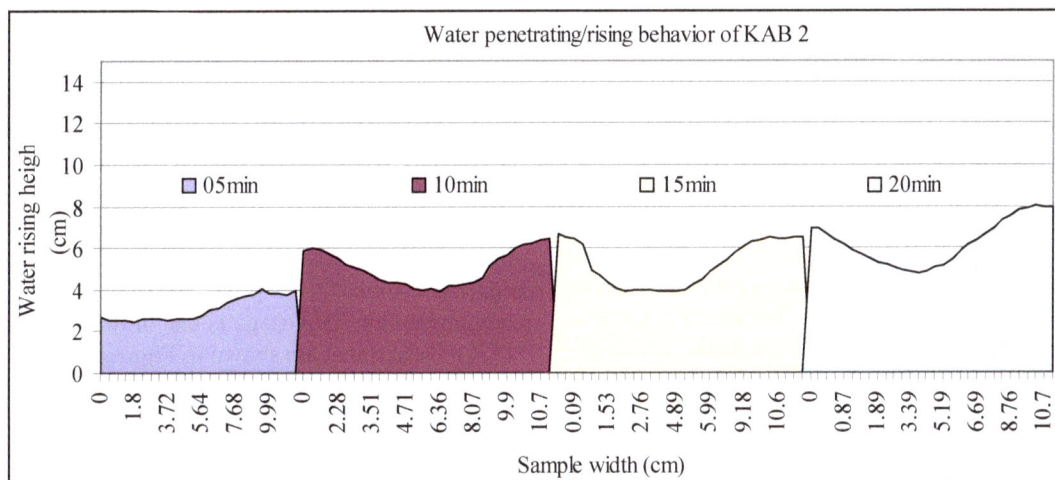

Figure 5. Water penetrating height at different immersion time for KAB 2.

Water penetrating height through the middle zone for KAB1 sample at 5, 10, 15 and 20 minute is 5 cm, 8.1 cm, 8.4 cm and 10.6 cm (**Figure 6**), respectively. In that case the water rising through the two edges and the middle is almost same for individual immersion case. The relation of incremental intrusion area which is indicated in the **Figures 2** and **3** of neutron radiographic images of the KAB samples at different immersion time is directly related to the IRA.

5.3. Determination of Incremental Intrusion Zone and Black/Gray Area

In the **Figures 2** and **3** is indicating the incremental intrusion zone *i.e.*, water is entering into a zone/place without encroachment and also shows the blue straight line. This blue line separates the actual immersion zone and the incremental intrusion zone of the immersed samples. Lower zone of the blue line is the actual immersion zone and the upper zone indicates the incremental im-

mersion zone of the immersed samples. After irradiation of the test (wet) samples KAB 1 & KAB 2, obtained the radiographic images of the wet samples by following the procedure (f) cited in the experimental part on neutron radiographic/Agfa structurix D4DW film as a latent image using neutron radiography method. For visualize this image it is transferred to the PC using high resolution camera and is viewed in the computer screen by the image analysis software Image J. With the help of this software, the total pixel distance corresponds to the total breath/length/height of the sample is calculated along x/y-axis. From that measurement, the number of pixels/cm breath or length or height of the sample is found. In the present investigation, the actual water length in a pan is 2 cm. So, by subtracting the actual water absorption zone (height, 2 cm) from the total water absorption zone of the immersed sample, incremental intrusion zone is found. This subtraction is done by the image analysis software. But, black area and gray area can clearly be

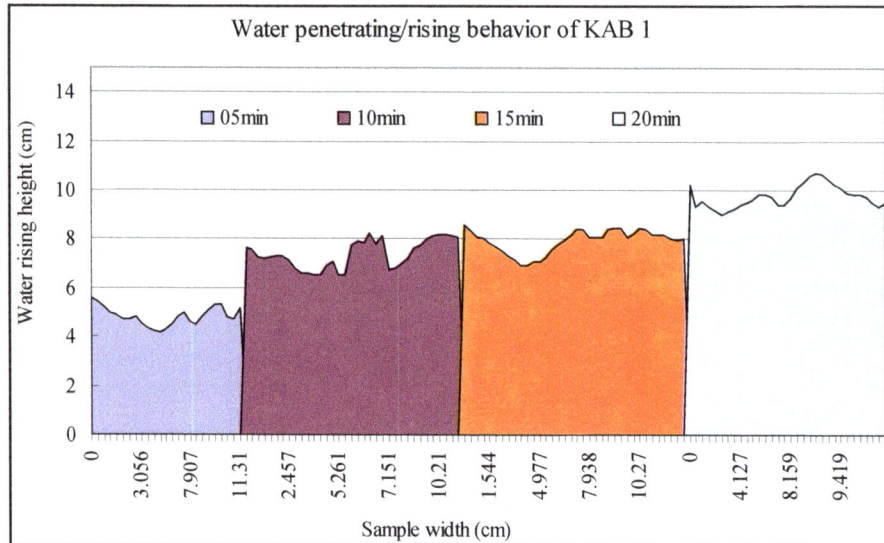

Figure 6. Water penetrating height at different immersion time for KAB 1.

distinguished only by taking radiographic images of the test sample with the help of neutron radiography (NR) method and the image analysis software. Because, Neutron radiography is a process of making a picture of the internal details of an object by the selective absorption of a neutron beam by the object and is a very efficient tool to enhance investigations in the field of non-destructive testing (NDT) as well as in many fundamental research applications. On the other hand, it is suitable for a number of tasks and impossible for conventional X-ray radiography. The advantage of neutrons compared to X-rays is the ability to image light elements (*i.e.* with low atomic numbers) such as hydrogen, water, carbon etc and can be distinguished gray area/black area of the radiographic image of the sample taken by the neutron radiography method.

5.4. Water Penetrating Behavior

Weng *et al.* studied that water absorption decreased significantly when the temperature increased due to the formation of the amorphous phase at high firing temperature. During the manufacturing time if the clay mixture absorbs more water, brick exhibits a larger pore size, resulting in a lower density. Depending on the H_2O absorption time of brick, observe differences in capillary absorption [36]. From the present investigation it also shows that the water rising/penetrating behavior through the different brick samples is like as stair, capillary, wave, zigzag shape. The resulting shape of the penetrating water into the different brick sample is shown in **Figures 5** and **6**.

5.5. Initial Rapid Absorption (IRA)

It is the measurement of the absorption rate that water is absorbed by a porous solid. It is related to the durability, porosity, pore size distribution and water absorption. It is sometimes called rising damp. The quantity, sizes and connection of pores influence the absorption rate of the brick. The IRA is reported in units of $g/(30 \text{ in}^2 \cdot \text{min})$ [37]. In the present case, IRA is measured in units of $gm/cm^3/min$. Robinson [37] described three stages of capillary absorption. IRA stage is one of them. The results of IRA measurement are shown in **Figure 7**.

In the case of KAB 2, the initial rapid absorption of water is less and for KAB 1 it is higher than KAB 2. At a glance the IRA for KAB brick sample can be written as KAB 2 < KAB 1. Low values of water absorption obtained in this study indicate that the clay bricks produced were poorly porous. Internal structure of the brick is expected to be intensive enough to avoid intrusion of water [36].

Dr. Robinson [37] found a relationship between capillarity and freeze thaw durability. He stated that durability is a function of the pore structure and the nature of the fired bond. On the other hand, capillary absorption measures how well water moves through the brick, then it must have some bearing on the efflorescence potential. Theoretically, the rate of capillary absorption influences the bond between brick and mortar (mixture of lime, water and sand). York dale did not believe that there was a direct relationship between IRA and performance and did not feel that IRA should be included in ASTM specifications. This disagreement is probably related to the lack of information contained in the IRA measurement. Workmanship plays such a large role in the quality of masonry that it is hard to definitively identify the influence of other factors. Rising damp and moisture transfer through masonry. For a particular type of brick which suggests

Quality Study of Automated Machine Made Environmentally Friendly Brick (KAB) Sample Using
Film Neutron Radiography Technique

147

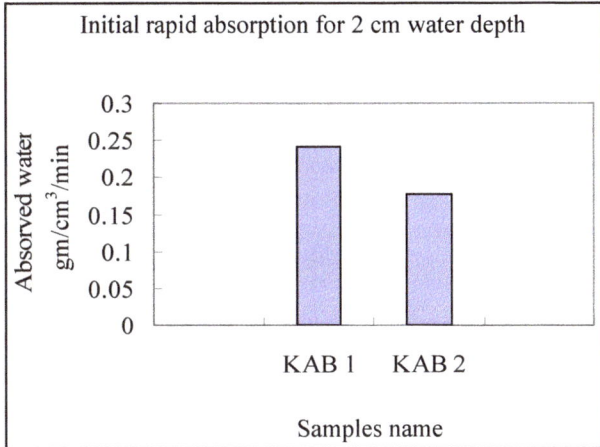

Figure 7. IRA measurement for KAB samples.

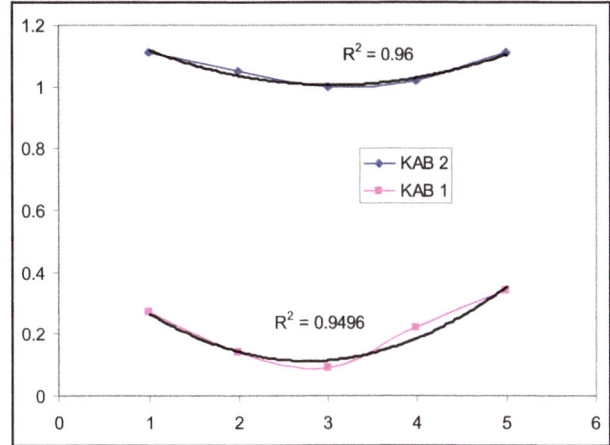

Figure 8. Optical densitometric measurement of different NR images.

that the connectivity and orientation of pores also play a large part in the movement of water in the pores [37].

6. Conclusions

The quality of a brick depends on the proper distribution of the contents/homogeneity, porosity, water penetrating behavior etc. in the sample. From the optical density measurement (**Figure 8**), it is observed that the optical density curve for KAB 2 sample is almost straight and for KAB 1, it shows far from straight line. From the points of optical density measurement, porosity, homogeneity, IRA and water penetrating behavior of view, it is pointed out that KAB is in good quality. The specific relationship between the shades of gray or density and exposure depends on the characteristics of the film emulsion and the processing conditions.

This absorption rate for KAB 2 is lower than that of KAB 1 and the water absorption increases with time gradually. **Figure 9** shows the water absorption characteristics of the samples. This indicates that after 5 minutes' immersion the absorption rate is very slow and becomes steady during long immersion time. In the case of KAB sample, steady time is higher. With higher steady time, slow absorption rate indicates the good quality. Many authors [38,39] studied that this absorption depends on submersion time, firing temperature and firing time. Few authors [40] investigate that when the mixture absorbs more water, brick exhibits a larger pore size, resulting in a lighter density.

7. Acknowledgements

Authors would like to thanks to the production manager, Kapita auto bricks limited, Joypura, Dhamrai, Dhaka, Bangladesh for supply the sample and to the Ministry of Science & Technology for their financial aid in order to collection/completion of this work.

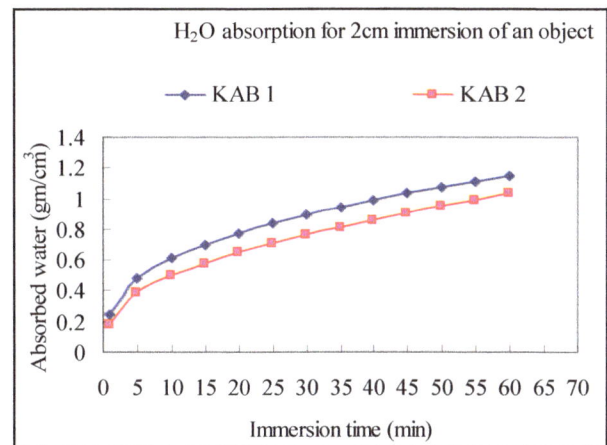

Figure 9. Water absorption rate of the sample (samples are immersed in water only 2 cm).

REFERENCES

[1] H. Berger, "Neutron Radiography," Elsevier, Amsterdam, 1964.

[2] P. Von der Hardt and H. Rotterger, "Neutron Radiography Hand Book," D. Reidel Publishing Company, Dordrecht, 1981.

[3] M. M. Rahman, S. Saha, M. N. Islam, M. K. Alam, A. K. M. A. Rahman and S. M. Azharul Islam, "A Study of the Morphological Change in Plant Pod by Using Neutron Radiography Technique," *Journal of African Review of Physics*, Vol. 8, 2013, pp. 239-242.

[4] J. M. Cimbala, D. E. Hughes, S. Levine and D. Sathianathan, "Application of Neutron Radiography for Fluid Flow Visualization," *Nuclear Technology*, Vol. 81, No. 3, 1988, pp. 435-445.

[5] M. N. Islam, M. M. Rahman, S. M. A. Islam and M. A. Zaman, "Neutron Radiographic Investigation of the Qual-

ity of Some Rubber Samples," *Indian Journal of Pure and Applied Physics*, Vol. 38, 2000, pp. 675-680.

[6]   M. K. Alam and M. A. Khan, "Study of Water Absorption and Internal Defects in Jute Reinforced Biopol Composite Using Digital Neutron Radiography Technique," *Journal of Bangladesh Academy of Sciences*, Vol. 30, No. 1, 2006, pp. 29-37.

[7]   C. Renfrew and P. Bahn, "Archaeology: Theories, Methods and Practice," Thames and Hudson, New York, 1996.

[8]   P. Rice, "Pottery Analysis: A Sourcebook," University of Chicago Press, Chicago, 1987.

[9]   M. K. Alam, "Comparative Study of Internal Defects in Ceramic Products Using CCD-Camera Based Digital Neutron Radiography Detector," *Bangladesh Journal of Scientific and Industrial Research*, Vol. 40, No. 3-4, 2005, p. 169.

[10]   M. K. Alam, M. N. Islam and M. A. Zaman, "Study of Internal Defects and Water Absorption Behavior of Single Layer Italian Tiles Using Neutron Radiography Facility of 3 MW TRIGA MARK II Research Reactor," *Journal of Bangladesh Academy of Sciences*, Vol. 31, No. 2, 2007, pp. 213-222.

[11]   M. N. Islam, M. K. Alam, M. A. Zaman, M. H. Ahsan and N. I. Molla, "Application of Neutron Radiography to Building Industries," *Indian Journal of Pure and Applied Physics*, Vol. 38, 2000, pp. 348-354.

[12]   M. K. Alam, M. A. Khan and E. H. Lehmann, "Comparative Study of Water Absorption Behavior in Biopol and Jute Reinforced Biopol Composite Using Neutron Radiography Technique," *Journal of Reinforced Plastics and Composites*, Vol. 25, No. 11, 2006, pp. 1179-1187.

[13]   M. N. Islam, M. A. Khan, M. K. Alam, M. A. Zaman and M. Matubayashi, "Study of Water Absorption Behavior in Wood Plastic Composite by Using Neutron Radiography Technique," *Polymer-Plastics Technology and Engineering*, Vol. 42, No. 5, 2003, pp. 925-934.

[14]   A. H. Bouma, "Methods for the Study of Sedimentary Structures," John Wiley and Sons, New York, 1969, p. 140.

[15]   P. M. Norris, J. S. Brenizer, D. A. Paine and D. A. Bostain, "Measurements of Water Deposition in Aerogel by Neutron Radiography," *Proceedings of 5th World Conference on Neutron Radiography*, Berlin, 17-20 June 1996, p. 602.

[16]   A. A. Harms and D. R. Wyman, "Mathematics and Physics of Neutron Radiography," D. Reidel Publishing Company, Holland, 1986, p. 22.

[17]   B. Illerhaus, "Proc. Seminar Computer Tomographie: Stand der Technik und Zukunftsaussichten," Family Readiness Group, Stuttgart, Vol. 29, 1988, p. 15.

[18]   G. Pfister, A. K. Schatz, C. Siegel, E. Steichele, W. Waschkowski and T. Bucherl, "Nondestructive Testing of Materials and Components by Computerized Tomography with Fast and Thermal Reactor Neutrons," *Nuclear Science and Engineering*, Vol. 110, No. 4, 1992, pp. 303-315.

[19]   ASTM-E545-05, "Standard Test Method for Determining Image Quality in Direct Thermal Neutron Radiographic Examination," 2010.
http://www.astm.org/Standards/E545.htm

[20]   H. P. Leeflang and J. F. W. Margraf, "Detection of Corrosion on Air Craft Components by Neutron Radiography," *Proceedings of the 4th World Conference on NR*, Sandiego, 10-16 May 1992, pp. 161-172.

[21]   S. Prasertsan and T. Theppaya, "A Study toward Energy Saving in Brick Making: Part 1-Key Parameters for Energy Saving," *International Energy Journal*, Vol. 17, No. 2, 1995, pp. 145-156.

[22]   J. M. Bhatnagar and R. K. Goel, "Thermal Changes in Clay Products from Alluvial Deposits of the Indo-Gangetic Plains," *Construction and Building Materials*, Vol. 16, No. 2, 2002, pp. 113-122.

[23]   G. Cultrone, E. Sebastian, K. Elert, M. J. Torre, O. Cazalla and C. Rodriguez-Navarro, "Influence of Mineralogy and Firing Temperature on the Porosity of Bricks," *Journal of the European Ceramic Society*, Vol. 24, No. 3, 2004, pp. 547-564.

[24]   L. Mbumbia, A. M. Wilmars and J. Tirlocq, "Performance Characteristics of Lateritic Soil Bricks Fired at Low Temperatures: A Case Study of Cameroon," *Construction and Building Materials*, Vol. 14, No. 3, 2000, pp. 121-131.

[25]   W. J. MC Burrey, "The Effect of Strength of Brick on Compressive Strength of Brick Masonary Process," Vol. 2, ASTM, West Conshohocken, 1970, p. 28.

[26]   "Manufacturing, Classification and Selection of Brick, Manufacturing Part I," Brick Industry Association, Reston, 1986.

[27]   O. Kayali and K. J. Shaw, "Manufactured Articles from Fly Ash," Patent no.PCT/AU03/01533, Australia, International Patent no. PCT/AU02/00593, European Patent Registration, R. Dhir, T. Dhir and J. Halliday, Eds., Publisher.

[28]   C. Renfrew and P. Bahn, "Archaeology: Theories, Methods and Practice," Thames and Hudson, New York, 1996.

[29]   P. Rice, "Pottery Analysis: A Sourcebook," University of Chicago Press, Chicago, 1987.

[30]   S. Bhatnagar and R. K. Goel, "Thermal Changes in Clay Products from Alluvial Deposits of the Indo-Gangetic Plains," *Construction and Building Materials*, Vol. 16, No. 2, 2002, pp. 113-122.

[31]   S. L. Marrusin, "Interior Fissures and Microstructure of Shale Brick," *American Ceramic Society Bulletin*, Vol. 64, No. 5, 1985, pp. 674-678. (Turkish Standards Institutions, Ankara, 1979).

[32]   M. A. Rahman, J. Podder and I. Kamal, "Neutron Radiography Facility in Bangladesh Research Reactor," *Proceedings of the 3rd World Conference on Neutron Radiography (NR)*, Osaka, 14-18 May 1989, pp. 179-185.

Quality Study of Automated Machine Made Environmentally Friendly Brick (KAB) Sample Using
Film Neutron Radiography Technique

149

[33] M. N. Islam, M. M. Rahman, M. H. Ahsan, A. S. Mollah, M. M. Ahasan and M. A. Zaman, "A Study of Neutron Radiography Parameters at the Tangential Beamport of the 3 MW TRIGA Research Reactor of AERE, Savar," *Jahangirnagar University Journal of Science*, Vol. 19, 1995, p. 181-187.

[34] H. I. Bjelkhagen, "Silver-Halide Recording Materials," Springer Verlag, 1993, pp. 128-151.

[35] T. J. Collins. "Image J for Microscopy," *BioTechniques*, Vol. 43, No. S1, 2007, pp. S25-S30.

[36] C. H. Weng, D. F. Lin and P. C. Chiang, "Utilization of Sludge as Brick Materials," *Advances in Environmental Research*, Vol. 7, No. 3, 2003, pp. 679-685.

[37] J. Sanders and J. Frederic, "Capillary—A New Way to Report IRA (initial rate of absorption)," *Structural Clay Products Division Meeting*, The National Brick Research Center, Gettysburg, 2-4 May 2011, pp. 1-24.

[38] "Clay Bricks (Wall Tile), TS 704," Turkish Standards Institution, Ankara, 1979.

[39] "Solid Bricks and Vertically Perforated Bricks, TS 705," Turkish Standards Institution, Ankara, 1975.

[40] S. Karaman, S. Ersahin and H. Gunal, "Firing Temperature and Firing Time Influence on Mechanical and Physical Properties of Clay Bricks," *Journal of Scientific & Industrial Research*, Vol. 65, No. 2, 2006, pp. 153-159.

16

The Integrated 3D As-Built Representation of Underground MRT Construction Sites

Naai-Jung Shih[*], **Chia-Yu Lee, Tzu-Ying Chan, Shih-Cheng Tzen**

Department of Architecture, National Taiwan University of Science and Technology, Taipei, Taiwan, China.

ABSTRACT

This study facilitates the scalability of as-built data from an earlier street level to underground transportation sites from the life-cycle perspective of urban information maintenance. As-built 3D scans of a 6 km street were made at different time periods, and of 3 underground Mass Rapid Transit (MRT) stations under construction in Taipei. A scanned point cloud was used to create a Building Information Modeling (BIM) Level of Development (LOD) 500 as-built point cloud model, with which topographic utility data were integrated and the model quality was investigated. The complex underground models of the transportation stations are proofed to be in correct relative locations to the street entrances on ground level. In the future the 3D relationship around the station will facilitate new designs or excavations in the neighborhood urban environment.

Keywords: Point Cloud; 3D Scans; As-Built Model; Building Information Modeling (BIM); Level of Development (LOD); Mass Rapid Transit (MRT)

1. Introduction

Transportation systems are an important indicator of urban development. The systems are subject to consistent monitoring from a life-cycle point of view, and a process being able to reflect actual construction conditions is needed. To ensure an appropriate construction simulation, the preconstruction preparation includes programming, scheduleing, methods, emergency procedures, etc. While data are created in different stages, 4D simulation is a powerful tool for the evaluation of construction processes [1], in which both data and the construction process can be visualized, allowing the communication of this information between different parties. Nevertheless, the simulation has limitations in terms of defining actual occurrences at a site when a very complicated collection of activities and objects is presented. The complexity adds difficulties and uncertainties in creating corresponding digital representations of the data.

Point cloud models are as-built data, whose integration with old environmental data leads to a specific application in showing most current status of environment or in contrasting the changes. The model can also be presented in virtual world, in which virtual 3D city models are becoming more widely implemented by governments and city planning services, of which highly detailed 3D models that reflect the complexity of city objects and the interrelations are required [2,3]. Nowadays, city modeling has reached a new level of reality in which 3D point cloud models are created with rich geometric properties and rich details, which enable the clouds to integrate other city model types [4].

The concept of rich geometric data should be extended to new underground construction site by being capable of integrating with existing models at street level for update purposes. However, technical, policy, and institutional barriers are usually faced in integrating data from multiple state-based sources [5]. Same situation can occur to departments of a local government for spatial-referenced multiple land information databases. The data from all platforms need to be exchangeable for the best efficiency [6]. Based on shared data, system integration can be achieved to support of planning decision-making and facility management after construction. The concept of cross-sourcing virtual cities [7] should be promoted further to as-built data in a city scale, as to reflect the real content of an environment. In addition, 2D registration processes should be extended to cover 3D property registration [8], like the integration of topographic map and as-built 3D city models.

Monitoring the development of city infrastructure is an important task. Geospatial technique is used to monitor city infrastructure networks by, for example, mobile laser scanning [9]. The issues to be taken care of include the representation, identification, and segmentation of 3D urban objects. Although CityGML is a common information model for the representation of 3D urban objects, such as buildings, traffic infrastructure, water bodies [10], the presence of these subjects needs to be verified by as-built model prior to evaluation or simulation. Although high-complexity point clouds have been collected from airborne terrestrial LiDAR 3D for city modeling [11,12] with greater efficiency, underground site needs to scan and to register clouds from inside the basements or tunnels by regions.

Technologies for mapping the underworld (MTU) have been applied to the condition assessment of underground utilities of buried infrastructure [13]. Although the scans could not be made during the occurrence of water, natural gas, electricity, telecommunications and sewerage. The underground scan not only presents the relationship with outside world, but also comes with specific scan-related data application, like rock engineering [14]. With semi-underground openings available, the connection between inside and exterior can be well-established with long-term measurements [15].

Increasing need has been shown in generating real world facilities in virtual environment, involving different levels of balance between human and computer effort [16]. With the balance in mind, after the environmental data are retrieved, the human effort is still needed especially in identifying the difference between heterogeneous representations among objects by initializing planar or cylinder shapes into walls, floors, ceilings, and pipes.

1.1. Research Scope

This study combines two types of as-built records, existing street facades and new underground construction, to extend the scope of present data and to set up a checkpoint for future data comparison. As-built records, which are used to monitor the quality of transportation systems, are usually difficult to create seamlessly between different phases, such as programming, design, simulation, construction, maintenance, and afterward. In order to determine the differences via comparison, new scanned data are registered with existing ones to define their interrelationships.

This independent scan project retrieved new Level of Development (LOD) 500 as-built models of underground MRT stations without LOD 100-400 data provided. The underground data are integrated into the as-built building point cloud models above ground level to extend existing LOD 500 data for future designs and excavations nearby. In order to facilitate greater integration with other disciplines [17], this is also considered as data collaboration from heterogeneous departments toward a finalized Building Information Modeling (BIM) model.

An LOD 500 at urban scale should be conducted prior to construction in order to facilitate any new design-related activity occurring in a neighborhood area with a broader evaluation perspective. Most urban scans are visualization-oriented, despite the result actually being a collection of single buildings at LOD 500 level. Since design and nearby environment are mutually influenced, the LOD 500 of nearby buildings should be required for an overall evaluation [18]. The related data are the configuration which contributes to the proportion, skyline, or orientation of the entire region. The most straightforward way to collect data is to scan and to examine the configuration based on as-built shape.

1.2. Methodology

This study recursively creates as-built representation for future reference. The as-built representation comes with different approaches, such as from modification from former design models based on field measurements. However, the complexity of building environment usually excludes the possibility of thorough data retrieval. Additionally, the accuracy check can be difficult for cross-referenced urban environment. In contrast to correcting building data from different departments, it's more important to verify individual data set and to create cross-reference among the sets.

2. MRT Stations and Scans

The East-West MRT line of Taipei, Taiwan, is separated into 7 sectors with different construction contractors and progress (**Figure 1**). Most of the excavation has been made underground, with connections to ground level through openings for the access of machinery, materials, or workers. The openings are usually located in the middle of streets carrying heavy traffic, and are fenced off with different arrangements of materials on the ground level.

Two scan sessions, one above ground level and one underground, were conducted at two different time periods. The former has registration points set up, and can be seen without visual interference. The latter has very limited area for registration. The combination shows the scalability of as-built data from an earlier and smaller

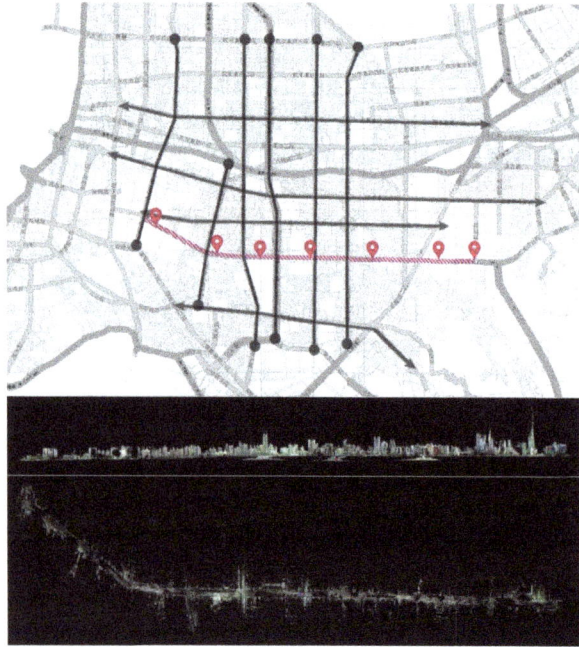

Figure 1. The newest East-West MRT line of Taipei and the point cloud model of the street.

amount of data to a broader life-cycle perspective of data. The scans were made by a Leica HDS 3000™ long-range laser scanner, which applies time-of-flight technology to calculate distance.

In total, scans were made at 59 locations (ScanWorld) with 1344 individual scans of different sizes. The entire scans took 21 days (not including preliminary visits, site meetings, planning), in which 37 ScanWorlds were deployed above ground and 22 ScanWorlds were made underground. A ScanWorld is the internal data representation of the scan database for locations: a ScanWorld may consist of many scans. The 1344 detail scans were made for registrations. In this study, a ScanWorld is usually made of a large area scan (up to 360×270 degrees) and a number of high resolution scans of features points as detail scans. With the point spacing of 10 - 20 cm at 100 m, it usually took about one hour for each ScanWorld and another half an hour for detail scans. The detail scans are important to the precision of final scan model and the following scan jobs, because it can create a correct of 3D spatial frame for future reference. The registration tolerance is about 8 - 12 mm/100 m.

In order to avoid any obstruction to a scan, the ground level was scanned from the roofs of nearby offices and apartments. The scan locations must be chosen in such a way as to avoid, or to cover, the blind spots near the bottom of the scanner. With a scan range of 250 meters, raising height actually broadens the covered ground area. The project chose scan locations at about every 100 - 150 meters. The scan process (**Figure 2**) is made of existing scans, new ground level scans, and the addition of un-

derground MRT scans, by referring to ground level scans.

The scan job is also divided into two parts: field scan and laboratory modeling. The former would need 3 - 4 persons for machine transportation, setup, and operation, the latter need only 2 - 3 persons for registration, data abstraction, modeling (point model, polygon model, rapid prototyping or RP model), and urban analysis (*i.e.* façade proportion, regional landscape).

After the excavation was completed, temporary covers were installed above the original street at the same location to store construction materials and fence panels. The locations and related point cloud can be seen in **Figure 3**. Each orange thin line represents the width of the point

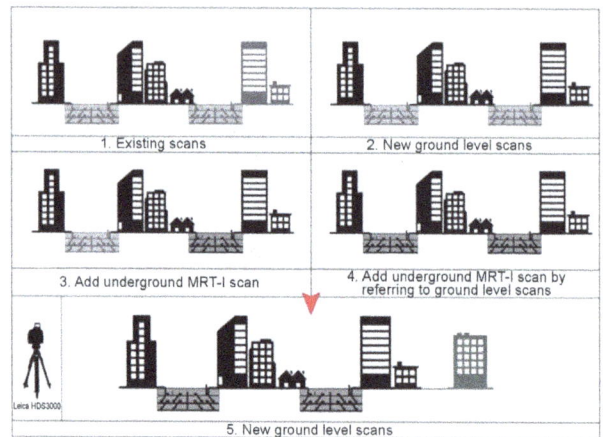

Figure 2. The scan sequence and references.

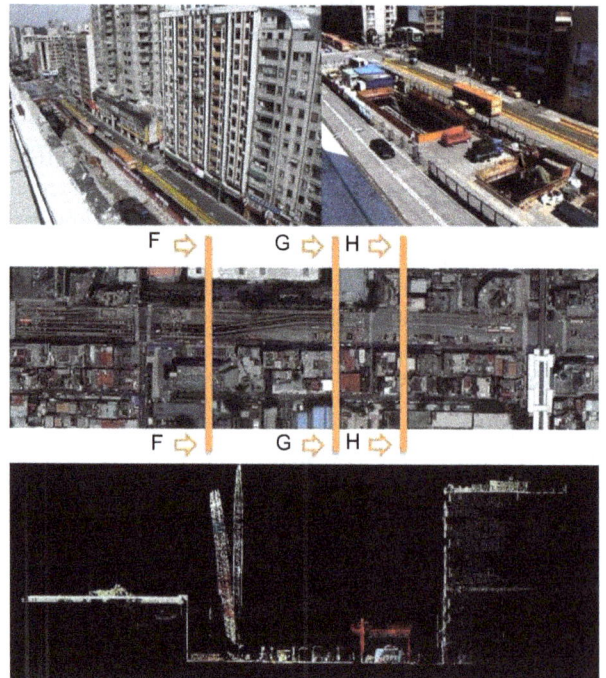

Figure 3. General construction scenes on ground level and the point cloud section at specific location G.

cloud that the section is made of, and the arrows indicate the viewing direction. The steel supports have not been completed, and the excavation, construction materials, and small machinery were placed on the ground level. Part of the construction under ground level was uncovered during this scan period.

Shin-I street is part of the main circulation system running East-West in Taipei. The MRT construction on street level and below has a significant impact on public transportation. The reduction of the original street width for excavation and temporary working decks intensifies traffic problems. To quantify the influence, the section at each interval is extracted to illustrate the percentage of construction over the entire street width for about 10% - 40%. Street profiles can vary according to buildings on both sides, and by construction-related activities. Large construction machinery, about two-stories high, is usually installed next to ground openings close to the middle of the street, dividing the space in half. The section profiles, which can also be seen in the point cloud (**Figure 3**), illustrate the narrow clearance between the machines and the facades.

2.1. Underground Construction of MRT Stations

The MRT stations and rails are constructed beneath one of the main streets in Taipei. Although this concealed infrastructure is connected to the daily activity above, the relative locations of the two are very important, as the complexity may affect future construction or renovations. Considering the types of complexity three-dimensionally, three MRT stations were studied as 1) Type A: a joint design and development with a park at ground level; 2) Type B: shared structure with the existing MRT station; and 3) Type C: a typical underground station. Types A and B are exemplified in the following sections.

Scans (Type A) were made at different locations, such as entrances, the ticket lobby and the platform (**Figure 4**). The scanned components included: structures, rails, tunnels, materials, and HVAC systems. Additional scans were made to combine the cloud models above ground level.

Retrieving sections at different locations of a cloud model helps in the comprehension of the construction process and the inter-relationships among components. The co-related component arrangement is more likely to uncover any missing interface between systems. The sections are directly made from the as-built cloud models; they are more likely to precisely react to construction errors. The cloud-derived plan (**Figure 5**) illustrates the locations of walls, columns, staircases and temporary storage areas. The platform can also be identified on the B2 level. The cross-section of the entire station shows the relative location of the steel structures of two floor levels and the temporary working platforms.

Figure 4. Point clouds of entrance, platform, rails and steel structure.

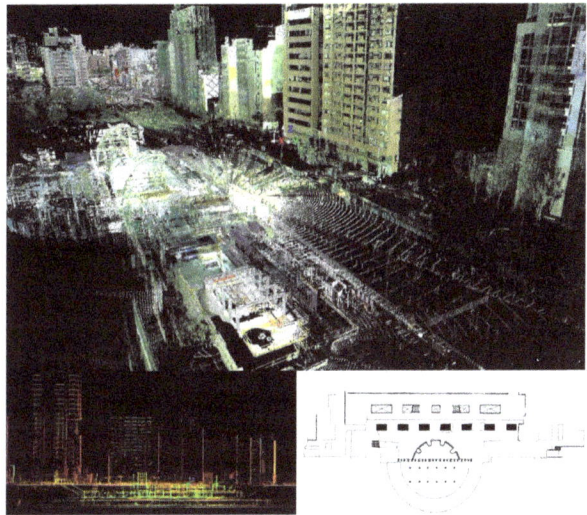

Figure 5. The point cloud of the urban environment at ground level and MRT underground (Type A).

2.2. A Top-Down Hierarchy of Cloud Models

A construction schedule consists of multiple concurrent or sequential activities. To record the entire perspective of the as-built 4D progress, the activities in terms of components have to be defined from, for example, temporary structures, excavation, foundation, steel bars, concrete, rails, and steel structure, to interior finishing. The complexity in terms of details is traditionally defined in a bottom-up structure in which each type has to relate its presence in regard to the entire perspective. To simulate the structure in this way would require tremendous effort. In contrast, a top-down viewpoint in defining complex construction activities uses the cloud model as the central database and subdivides each construction as needed at each schedule checkpoint for inspection. From the design point of view, a building is the product of a top-down process. Thus, a building cloud model is defined from a construction record point of view, which is useful because of its similar top-down nature. Most importantly, the cloud model is a feasible means of recovering the geometric construction conditions at a certain level, compared to the limited perspective of photographs, videos, 2D drawings, or 3D design models, in which the data segmentation characteristics can hardly be related to each other; the as-built database possibly would not even exist unless scans are made.

The point cloud can be navigated by users trained to read this specific type of data representation, and a project leader can also request the cloud be rotated, panned, or scaled, as needed, to look for a specific point as a vertex or a set of linear points as an edge for measurements.

3. Inspection of the Relationship between the MRT Station and the Urban Environment

The as-built facades and cross-sections of a street are very difficult to create. Government infrastructure surveys refer to traditional field data. With private property buildings, the data of an entire street or block are not only difficult to integrate as a whole, but access to the data of each private building is usually restricted. Current 2D drawings have limited block-wise information in both vertical and horizontal dimensions, especially when dealing with all street-facing buildings. Due to a lack of updated information on new and old buildings, as-built data become the only source of information for any new construction project.

3D scans of an entire block and street enable the creation of as-built data, which integrate not only various private buildings, but also the co-relationship between government and private sources of data. The integrated and co-related data enable the generation of various types of drawings [19], which eliminates the needs to visit a large number of parties, to handle the integration of sources, and to verify the tolerance of measurements. The integrated data are presented with colors that create a more effective visualization of a larger area with a scan precision of up to 4 mm/50 m (Leica HDS 3000™). An end user not only has a larger perspective of a certain region, but also the ability to use the data by simply requesting a part of the point cloud at a specific location, and the traditional sections or elevations can be created with consistency.

The Type B MRT station is located in a transportation building shared by two lines: one above ground, which was built earlier, and another underground, which was under construction. The grey part of the point cloud model was created before 2010.7.15 above ground level, and the color part of the underground model was created before 2011.3.24 (**Figure 6**). With a comparison of the section with/out the new construction (**Figure 7**), the station and its entrances can be correlated to the existing urban environment outside the station. The vertical layout of the basement levels, corridors and exits can also be seen, with their alignment to the building entrances on the ground level.

Streets, buildings, and landscape above ground level can be seen and retrieved from 2D drawings and photographs. However, the relationship between the urban environment on the ground level and the underground MRT station is not easily discernible. In order to address this problem, a cloud model, which combines data from the ground level and above, can be oriented to fulfill inspection needs. For example, the relative location between the projection of the station and the North section of the street is shown in **Figure 6**; it can be seen that the station is located right below the street, and the distance between the station and the street buildings can be determined. The cloud-derived section is used to precisely measure the building heights and the depth of the station. The 3D relationships around the station will be used to accommodate new designs and nearby excavation.

4. Point Cloud for BIM

BIM consists of 3D models, and is used for qualitative and quantitative estimation. Each building object is defined as an element, and is subject to a LOD [20]. Both the BIM and point cloud models (PCM) inherit the

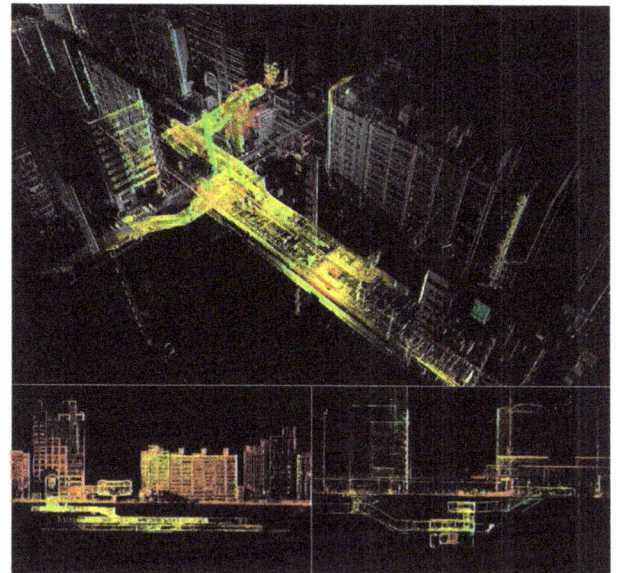

Figure 6. The point cloud of an MRT underground station (Type B) with the connections to ground level, the existing urban environment on the ground level are shown in gray color.

Figure 7. MRT with/out the new construction.

chronological nature of time, in which the former presents a forward design process, while the latter presents a reversed construction checking procedure, *i.e.* as the real physical objects are constructed, the scanned shape which presents as-built data is used to modify or check the LOD 300/400 for the final LOD 500 model.

BIM with object attributes can record detailed information for each building component. Although a programming language (like AutoLisp) can set definitions for vector drawing data, a user would need programming experience, skills possessed by a small number of designers or draftspersons. As with the component-based BIM, this study emphasizes an urban environment which is made of individual buildings as components.

The definition of 500 data should specify its nature in terms of execution process and the final result, and address the 4D characteristics [21] in an as-built manner. Although design and construction have been successfully defined or conducted a virtual environment, the design model has to be confirmed with as-built model. The LOD 500 usually comes after the final construction stages when most of the components are sealed or covered by finishes, and the confirmation of dimensions or shapes is usually prevented. In order to carry out 500 data, the 3D scans should be performed on as-built parts throughout the construction process [22] and monitor as-built data in terms of dimensions, configuration and adjacency [23].

This study creates a point-cloud-based data by the following steps:
- Separate cloud by parts
- Import to Autodesk Revit^{TM}
- Create 3D models
- Export images and models

The cloud data were sliced into plans, elevations and sections to be exported for modeling collaboration. Using Leica CloudWorx^{TM}, the cloud data were imported to Revit^{TM} (**Figure 8**). In order to increase modeling speed, the cloud data were separated by plans, elevations and sections as appropriate parts to be distributed for concurrent model making by multiple persons. The 3D cloud data were in full scale, which could be measured directly. The 3D data were also presented with 2D images as backgrounds to trace model boundaries.

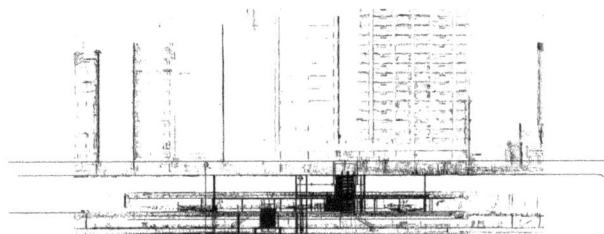

Figure 8. The type A MRT station cloud model is imported into Autodesk Revit^{TM}.

4.1. Creating 3D Models

Traditional modeling result confirms the design-oriented definition better than with construction-oriented as-built data. Complicated situations usually occur to a well-defined BIM, because quality control problems may arise in any phase of construction. A 3D as-built model can be used to verify the BIM data during construction, and to create a final 3D model after construction is completed. Cloud data clearly specify the dimensions of interiors and exteriors. Building components, such as beams, columns, walls, floors, staircases, etc., can also be identified. Scans can be applied before entities installed to inspect clearance around.

Overlaying construction (or fabrication) models and cloud models constitutes the most straightforward method of quality control in terms of checking boundaries, locations, clearances, or offsets by distances or regions. The related inspection can be viewed from all angles, and the model data can be sliced as sections at preferred intervals to avoid viewing obstructions. Most importantly, the level of construction accuracy can be justified.

The BIM for an HVAC system includes the ducts, joints, pipes and supporting accessories with specified clearances to ceiling or decks. Since the installation is based on an approved design, no intersection between components is expected in the as-built cloud models. However, the confirmation of the diameters and slopes of pipes that cross large spans, or have areas obscured by other components or partitions, are difficult to measure. If the sag of these components along linear paths is a few millimeters or inches, the deviation is usually trivial and does not show up in a BIM fabrication model. As a result, the final locations can never be determined when the tolerances differ from the designed specifications. The missing verification of the final dimensions of a component's size and location will transfer the tolerance to the following stage in facility management. Thus, the actual state of a component cannot be determined, especially when a component is sealed inside a piece of concrete wall or behind a fixed partition.

Scans were made of the MRT station before interior finishing was completed. These scans were able to record the HVAC system before it was covered by ceiling tiles (**Figure 9**). Cloud models can be used to create final drawings in the traditional way, or they can be used for confirmation with BIM. The point clouds are also used to estimate diameters. As shown in **Figure 10** (top), the cloud slice is imported into AutoCAD^{TM}. The locations and diameters of pipes are retrieved after the pipe-related points are initialized as 3D tubes. The scanned point clouds were used to create as-built polygon model at MRT lobby level (**Figure 11**).

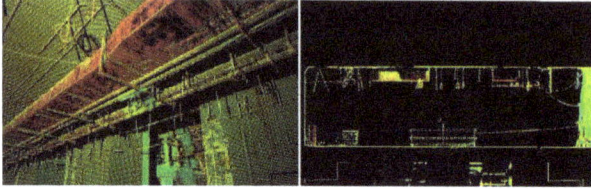

Figure 9. Section showing ducts, pipes, accessories, and partially finished ceiling.

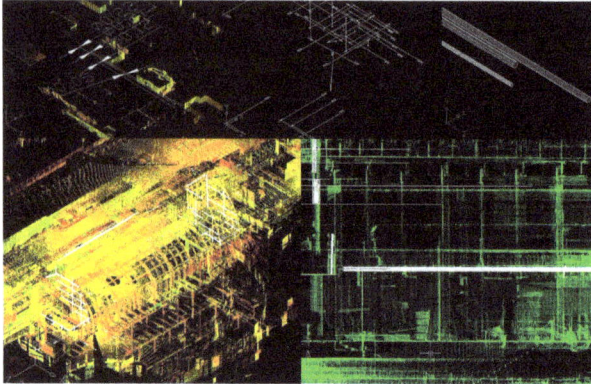

Figure 10. The floor and the retrieval of pipes (top); the arrangement and the overlapping with the cloud model (bottom).

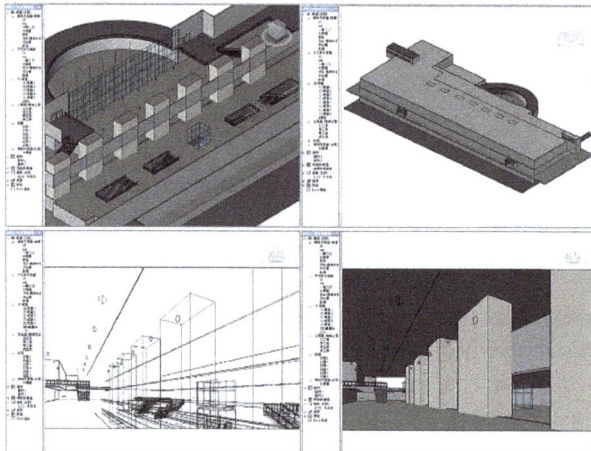

Figure 11. The polygon model for a type A MRT station based on as-built scans.

4.2. Export Images and Models

Once Autodesk Revit™ has used to create the BIM data, the information can be exported as 3D models, drawings, or images, and can be used in other applications or browsing software. This study imports 3D models into Geomagic Studio Qualify™ to compare the deviation between the point cloud model and polygon model (**Figure 12**). Construction companies can confirm quality control by first scanning, and then overlapping the scans with the LOD 300/400 model for possible misalignment. Any occurrence during the construction process in which

the as-built data do not align, indicates a possible problem as an offset from the original BIM representation.

The hardware burden in this project was reduced by avoiding the addition of too much information in a single file. Although BIM comes with 3D information, the model can still combine bitmap images or vector maps for the purpose of integrating building and urban data The AutoCAD™-exported MRT model (.dxf) is 19,478 KB, in contrast to the size of vector drawings of 2745 KB, with a larger area covered in a smaller size. It only requires 54.9 MB (12.3%) of RAM.

The 3D cloud model is also used to cross-reference values and to integrate information (**Figure 13**). For existing facilities, underground pipes, street lamps and fire hydrants, the construction process can be delayed unexpectedly. BIM model can integrate various maps as GIS to facilitate M&E execution.

Scan data can also meet traditional needs in architectural practice. For example, working drawings are usually created before the construction stage. The drawings specify quality control by measurements. Nevertheless, reference to after-construction structures can be difficult if barriers are obstructing the area to be measured. This problem is solved by referring to the cloud model, and by either directly measuring or using editing tools to remove the interference. The sections are similar to plans, except different projection angles can be selected immediately by slicing corresponding parts.

4.3. BIM Problem of the Building under Construction

The problems involved in directly applying point cloud data for BIM checking include the following:
- The surface model file created from the point cloud is too large to be easily manipulated. To manipulate large data set inside scan software is efficient because better imbed algorithm is provided for fast browsing and editing. An easy and straightforward way to accelerate the manipulation is to increase computer power and reduce cloud model size by boxing the needed part only for domain specific data application.
- Scan data may be insufficient or incomplete due to the viewing angles being blocked by objects. Additional scans have to be made. Scan from different angles have to be planned to recover the missing part of geometries.
- Scans only record surface details, and the internal composition of some components cannot be known. One way to know the internal composition is to scan ahead of the construction schedule, before the components being sealed.
- Finishing is incomplete during this project period, which leads to the final surface smoothness and the construction quality level being unknown. Building

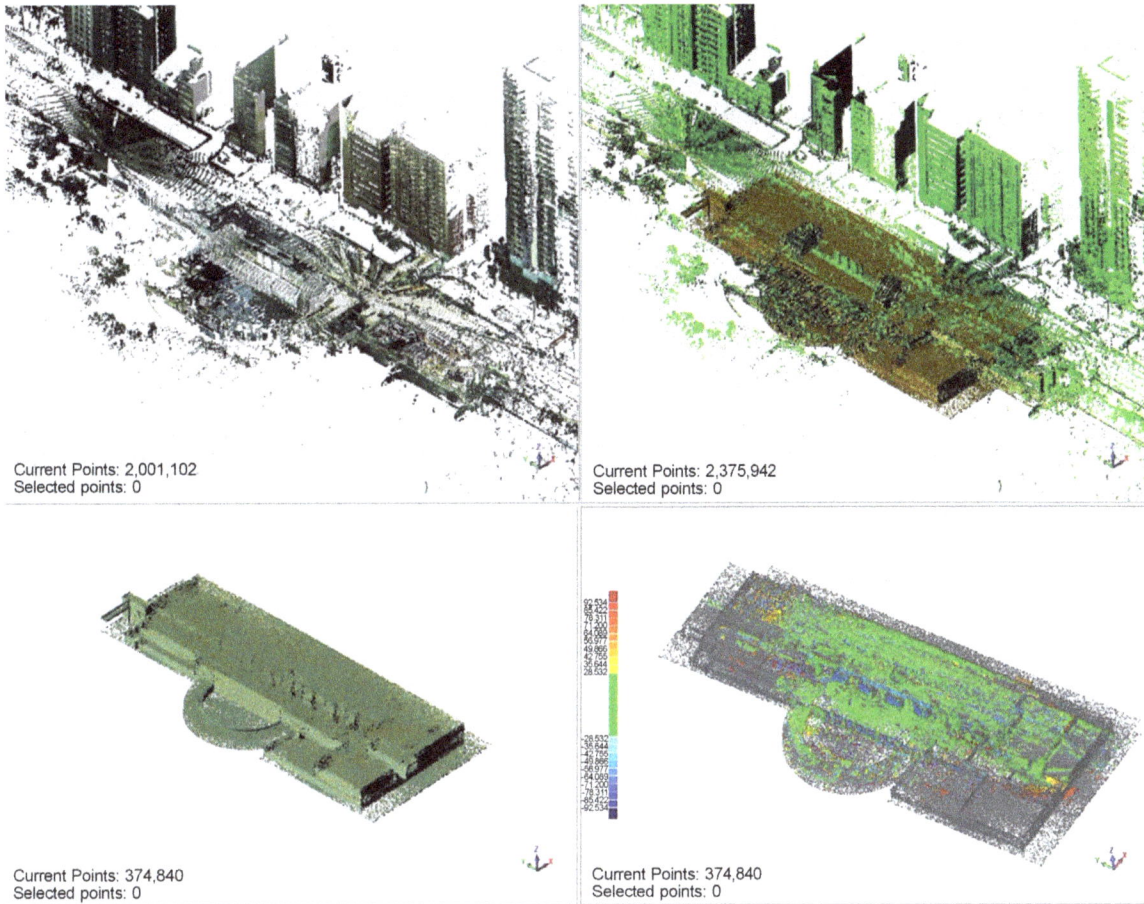

Figure 12. The combination of the cloud model (top left) and the polygon model (bottom left) into the alignment check of the main platform (bottom right).

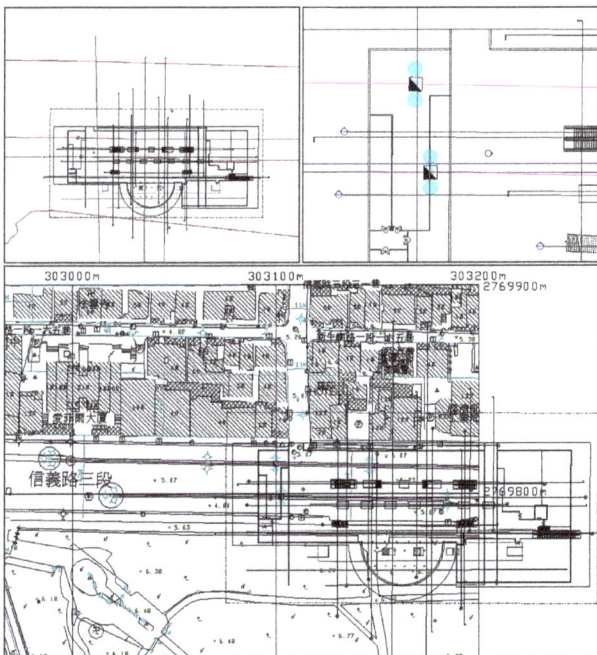

Figure 13. Overlapping BIM model with the maps of street lamps, utility boxes, pipes, and topographic information.

life cycle consists of different phases. Scans should be made accordingly for thorough records. In the future, scans need to be made again after building is occupied for a while when budget is available.

- Scans cannot be applied to transparent or reflective materials which would need other auxiliary field measurement device (total station or tape measure) for data retrieval.

5. Conclusions

This project was restricted by resources, and it was unable to produce a thorough life-cycle record of all stages. Although this MRT line was in the final stages of construction and is running just before the end of 2013, a future study could include issues based on the data created in this project and data integration pattern for new constructions. Chronological scans should be conducted after interior finishing and at least one year after the start of operation for post-occupation response. As stated in the methodology, the recursively defined as-built representation and related framework will contribute to future nearby construction works, and lead to a better start in

BIM since this project.

In order to combine the cloud model and original BIM in the design stage, this project was conducted as a local pioneer study, without any access to the internal MRT data, which was withheld for security reasons. Although no evaluation with the existing model was made, the project did create drawings and share the cloud model with the MRT administration. As a result, the project was conducted as an independent source, which is feasible for quality and schedule control.

6. Acknowledgements

The authors would like to thank the MRT administration for the scan-related assistance.

This project was sponsored by the National Science Council, Taiwan, the Republic of China, under the project number 98-2221-E-011-123-MY3. The authors would like to thank the Council for its support.

REFERENCES

[1] M. Fischer and J. Kunz, "The Scope and Role of Information Technology in Construction," CIFE Technical Report #156, Stanford University, Stanford, 2004.

[2] I. Jazayeri, "Chapter 8: Trends in 3D Land Information Collection and Management," In: A. Rajabifard, I. Williamson and M. Kalantari, Eds., *A National Infrastructure for Managing Land Information—Research Snapshot*, The University of Melbourne, Parkville, 2012, pp. 81-87.

[3] R. Stadler and T. H. Kolbe, "Spatio-Semanti Coherence in the Integration of 3D City Models," Commission II, WG II/7, 2013. http://130.203.133.150/viewdoc/summary;jsessionid=77B46D32CB56462266136F4EB89ECD9C?doi=10.1.1.221.7921

[4] S. Nebiker, S. Bleisch and M. Christen, "Rich Point Clouds in Virtual Globes—A New Paradigm in City Modeling?" *Computers, Environment and Urban Systems*, Vol. 34, No. 6, 2010, pp. 508-517.

[5] A. Rajabifard, I. Williamson and M. Kalantari, "A National Infrastructure for Managing Land Information—Research Snapshot," The University of Melbourne, Parkville, 2012.

[6] I. D. Bishop, F. J. Escobar, S. Karuppannan, K. Suwarnarat, I. P. Williamson, P. M. Yates and H. W. Yaqub, "Spatial Data Infrastructures for Cities in Developing Countries: Lessons from the Bangkok Experience," *Cities*, Vol. 17, No. 2, 2000, pp. 85-96.

[7] M. Uden and A. Zipf, "Open Building Models: Towards a Platform for Crowdsourcing Virtual 3D Cities," In: *Progress and New Trends in 3D Geoinformation Sciences, Lecture Notes in Geoinformation and Cartography*, Springer, Berlin, 2013, pp. 299-314.

[8] D. Shojaei, "Chapter 9: 3D Visualisation as a Tool to Facilitate Managing Land and Properties," In: A. Rajabifard, I. Williamson and M. Kalantari, Eds., *A National Infrastructure for Managing Land Information—Research Snapshot*, The University of Melbourne, Melbourne, 2012, pp. 88-94.

[9] M. Al-Hader, A. Rodzi, A. R. Sharif and N. Ahmad, "Mobile Laser Scanning for Monitoring Polyethylene City Infrastructure Networks," *Journal of Geography and Regional Planning*, Vol. 4, No. 6, 2011, pp. 364-370.

[10] B. Mao, "Visualisation and Generalisation of 3D City Models," Ph.D. Dissertation, Royal Institute of Technology, Stockholm, 2011.

[11] J. Heo, S. Jeong, H. K. Park, J. Jung, S. Han, S. Hong and H. G. Sohn, "Productive High-Complexity 3D City Modeling with Point Clouds Collected from Terrestrial LiDAR," *Computers, Environment and Urban Systems*, Vol. 41, 2013, pp. 26-38.

[12] A. Jochem, B. Hofle, V. Wichmann, M. Rutzinger and A. Zipf, "Area-Wide Roof Plane Segmentation in Airborne LiDAR Point Clouds," *Computers, Environment and Urban Systems*, Vol. 36, No. 1, 2012, pp. 54-64.

[13] T. Hao, C. D. F. Rogers, N. Metje, D. N. Chapman, J. M. Muggleton, K.Y. Foo, P. Wang, S. R. Pennock, P. R. Atkins, S. G. Swingler, J. Parker, S. B. Costello, M. P. N. Burrow, J. H. Anspach, R. J. Armitage, A. G. Cohn, K. Goddard, P. L. Lewin, G. Orlando, M. A. Redfern, A. C. D. Royal and A. J. Saul, "Condition Assessment of the Buried Utility Service Infrastructure," *Tunnelling and Underground Space Technology*, Vol. 28, 2012, pp. 331-344.

[14] O. Aydana and R. Ulusay, "Geotechnical and Geo-environmental Characteristics of Man-Made Underground Structures in Cappadocia, Turkey," *Engineering Geology*, Vol. 69, No. 3-4, 2003, pp. 245-272.

[15] Ö. Kaşmer, R. Ulusay and M. Geniş, "Assessments on the Stability of Natural Slopes Prone to Toe Erosion, and Man-Made Historical Semi-Underground Openings Carved in Soft Tuffs at Zelve Open-Air Museum (Cappadocia, Turkey)," *Engineering Geology*, Vol. 158, 2013, pp. 135-158.

[16] M. Fumarola and R. Poelman, "Generating Virtual Environments of Real World Facilities: Discussing Four Different Approaches," *Automation in Construction*, Vol. 20, No. 3, 2011, pp. 263-269.

[17] Y. Arayici, P. Coates, L. J. Koskela, M. Kagioglou, C. Usher and K. O'Reilly, "BIM Adoption and Implementation for Architectural Practices," *Structural Survey*, Vol. 29, No. 1, 2011, pp. 7-25.

[18] C. Eastman, P. Teicholz, R. Sacks and K. Liston, "BIM Handbook—A Guide to Building Information Modeling for Owners, Managers, Designers, and Contractors," John Wiley & Sons, Inc., Hoboken, 2008.

[19] N. J. Shih, "A Study of 2D- and 3D-Oriented Architec-

tural Drawing Production Methods," *Automation in Construction*, Vol. 5, No. 4, 1996, pp. 273-283.

[20] American Institute of Architects, "AIA Document E202™-2008 Building Information Modeling Protocol Exhibit," American Institute of Architects, Washington DC, 2008.

[21] W. Kymmell, "Building Information Modeling: Planning and Managing Construction Projects with 4D CAD and Simulations," McGraw Hill Construction, New York, 2008.

[22] N. J. Shih and S. T. Huang, "3D Scan Information Management System (3DSIMS) for Construction Management," *Journal of Construction Engineering and Management*, Vol. 132, No. 2, 2006, pp. 134-142.

[23] N. J. Shih and P. H. Wang, "Point-Cloud-Based Comparison between Construction Schedule and As-Built Progress—A Long-Range 3D Laser Scanner's Approach," *Journal of Architectural Engineering*, Vol. 10, No. 3, 2004, pp. 98-102.

Permissions

The contributors of this book come from diverse backgrounds, making this book a truly international effort. This book will bring forth new frontiers with its revolutionizing research information and detailed analysis of the nascent developments around the world.

We would like to thank all the contributing authors for lending their expertise to make the book truly unique. They have played a crucial role in the development of this book. Without their invaluable contributions this book wouldn't have been possible. They have made vital efforts to compile up to date information on the varied aspects of this subject to make this book a valuable addition to the collection of many professionals and students.

This book was conceptualized with the vision of imparting up-to-date information and advanced data in this field. To ensure the same, a matchless editorial board was set up. Every individual on the board went through rigorous rounds of assessment to prove their worth. After which they invested a large part of their time researching and compiling the most relevant data for our readers. Conferences and sessions were held from time to time between the editorial board and the contributing authors to present the data in the most comprehensible form. The editorial team has worked tirelessly to provide valuable and valid information to help people across the globe.

Every chapter published in this book has been scrutinized by our experts. Their significance has been extensively debated. The topics covered herein carry significant findings which will fuel the growth of the discipline. They may even be implemented as practical applications or may be referred to as a beginning point for another development. Chapters in this book were first published by Scientific Research Publishing Inc.; hereby published with permission under the Creative Commons Attribution License or equivalent.

The editorial board has been involved in producing this book since its inception. They have spent rigorous hours researching and exploring the diverse topics which have resulted in the successful publishing of this book. They have passed on their knowledge of decades through this book. To expedite this challenging task, the publisher supported the team at every step. A small team of assistant editors was also appointed to further simplify the editing procedure and attain best results for the readers.

Our editorial team has been hand-picked from every corner of the world. Their multi-ethnicity adds dynamic inputs to the discussions which result in innovative outcomes. These outcomes are then further discussed with the researchers and contributors who give their valuable feedback and opinion regarding the same. The feedback is then collaborated with the researches and they are edited in a comprehensive manner to aid the understanding of the subject.

Apart from the editorial board, the designing team has also invested a significant amount of their time in understanding the subject and creating the most relevant covers. They scrutinized every image to scout for the most suitable representation of the subject and create an appropriate cover for the book.

The publishing team has been involved in this book since its early stages. They were actively engaged in every process, be it collecting the data, connecting with the contributors or procuring relevant information. The team has been an ardent support to the editorial, designing and production team. Their endless efforts to recruit the best for this project, has resulted in the accomplishment of this book. They are a veteran in the field of academics and their pool of knowledge is as vast as their experience in printing. Their expertise and guidance has proved useful at every step. Their uncompromising quality standards have made this book an exceptional effort. Their encouragement from time to time has been an inspiration for everyone.

The publisher and the editorial board hope that this book will prove to be a valuable piece of knowledge for researchers, students, practitioners and scholars across the globe.

List of Contributors

Gwang-Hee Kim and Yoonseok Shin
Department of Plant & Architectural Engineering, Kyonggi University, Suwon-Si, Korea

Jae-Min Shin
Department of Architectural Engineering, Graduate School of Kyonggi University, Suwon-Si, Korea

Sangyong Kim
School of Construction Management and Engineering, University of Reading, Reading, UK

Keun-Hyeok Yang
Department of Plant & Architectural Engineering, Kyonggi University, Suwon, Korea

Kyung-Ho Lee
Department of Architectural Engineering, Kyonggi University Graduate School, Seoul, Korea

Seok-Jin Kang
School of Architecture, Gyeongsang National University, Jinju-Si, Korea

Donghoon Lee, Manki Kim and Sunkuk Kim
Department of Architectural Engineering, Kyung Hee University, Seoul, Republic of Korea

Robert E. Steffen, Sung Joon Suk, Yong Han Ahn and George Ford
Department of Construction Management, Western Carolina University, Cullowhee, USA

Zhen Zhang, Woo-Hwan Lee, Young-Wha Choi and Sung-Hoon An
Department of Architectural Engineering, Daegu University, Gyeongsan-Si, South Korea

Md. Zakaria Hossain
Department of Environmental Science and Technology, Graduate School of Bioresources, Mie University, Tsu, Japan

Boong Yeol Ryoo and Mike T. Duff
Department of Construction Science, Texas A&M University, College Station, USA

Kyung Wook Seo
Department of Architecture, Kyonggi University, Suwon, South Korea

Chang Sung Kim
Department of Architectural Engineering, Hyupsung University, Hwaseong, South Korea

Joonsuk Ahn
Department of Architecture, Kyonggi University, Suwon, South Korea

Junli Yang and Ibuchim Cyril B. Ogunkah
Department of Construction, School of Architecture and the Built Environment, University of Westminster, London, UK

Kui Zhao
School of Architecture & Urban Planning, Huazhong University of Science and Technology, Wuhan, China

William L. Tilson
School of Architecture, University of Florida, Gainesville, USA

Dan Zhu
Urban and Regional Planning, University of Florida, Gainesville, USA

Khurshed Alam, Sudipta Saha and Nurul Islam
Institute of Nuclear Science and Technology, AERE, Savar, Bangladesh

Syed Azharul Islam and Robiul Islam
Department of Physics, Jahangirnagar University, Savar, Bangladesh

Naai-Jung Shih, Chia-Yu Lee, Tzu-Ying Chan and Shih-Cheng Tzen
Department of Architecture, National Taiwan University of Science and Technology, Taipei, Taiwan, China

www.ingramcontent.com/pod-product-compliance
Lightning Source LLC
Chambersburg PA
CBHW070245230326
41458CB00099B/5257